Gender, Migration and Social Transformation

Intersectionality can be used to analyse whether migration leads to changes in gender relations. This book finds out how migrants from a peri-urban neighbourhood on the outskirts of Cochabamba, Bolivia, make sense of the migration journeys they have undertaken.

Migration is intrinsically related to social transformation. Through life stories and community surveys, the author explores how gender, class, and ethnicity intersect in people's attempts to make the most of the opportunities presented to them in distant labour markets. While aiming to improve their economic and material conditions, migrants have created a new transnational community that has undergone significant changes in the ways in which gender relations are organised. Women went from being mainly housewives to taking on the role of the family's breadwinner in a matter of just one decade.

This book asks and addresses important questions such as: what does this mean for gender equality and women's empowerment? Can we talk of migration being emancipatory? Does intersectionality shed light in the analysis of everyday social transformations in contexts of transnational migrations? This book will be useful to researchers and students of human geography, development studies and Latin America area studies.

Tanja Bastia is a Senior Lecturer at the Global Development Institute at the University of Manchester, UK.

Gender, Space and Society

The series on Gender, Space and Society publishes innovative feminist work that analyses men's and women's lives from a perspective that exposes and is committed to challenging social inequalities and injustices. The series reflects the ongoing significance and changing forms of gender, and of feminist ideas, in diverse social, geographical and political settings and adopts innovative methodological and philosophical approaches to understanding gender, space and society.

Series Editors:
Peter Hopkins, Newcastle University, UK
Rachel Pain, University of Durham, UK

The Illegal City
Space, Law and Gender in a Delhi Squatter Settlement
Ayona Datta

Masculinities and Place
Andrew Gorman-Murray and Peter Hopkins

Feminist Geopolitics
Material States
Deborah P. Dixon

Public Urban Space, Gender and Segregation
Women-only Parks in Iran
Reza Arjmand

Transforming Gender, Sex, Place, and Space
Gender Variant Geographies
Lynda Johnston

Gender, Migration and Social Transformation
Intersectionality in Bolivian Itinerant Migrations
Tanja Bastia

For more information about this series, please visit: www.routledge.com/Gender-Space-and-Society/book-series/ASHSER1355

Gender, Migration and Social Transformation
Intersectionality in Bolivian Itinerant Migrations

Tanja Bastia

LONDON AND NEW YORK

First published 2019 by Routledge

2 Park Square, Milton Park, Abingdon, Oxon, OX14 4RN
605 Third Avenue, New York, NY 10017

Routledge is an imprint of the Taylor & Francis Group, an informa business

First issued in paperback 2020

British Library Cataloguing-in-Publication Data
A catalogue record for this book is available from the British Library

Library of Congress Cataloging-in-Publication Data
A catalog record has been requested for this book

ISBN: 978-1-4724-3237-7 (hbk)
ISBN: 978-0-367-72823-6 (pbk)

Typeset in Times New Roman
by Taylor & Francis Books

Para Sahara y todas las otras migrantes que hacen del mundo su casa

Contents

List of illustrations   ix
Acknowledgements   xi
Glossary of Spanish terms   xiii

1   Introduction   1

2   Gender, migration, and social transformation   21

3   Origins   43

4   Mobility and social networks   63

5   Work   89

6   Care   113

7   Intimacy   137

8   Conclusion   155

References   163
Index   178

Illustrations

Figures

3.1 Year of arrival to the Barrio 52
3.2 Year of first migration, by sex 58

Images

1.1 Dancing the *diablada* at the annual neighbourhood festivity,
Cochabamba 5
1.2 View of the neighbourhood, Cochabamba 6
1.3 Informal settlement, City of Buenos Aires 16
1.4 One of the fieldwork sites, City of Buenos Aires 17
1.5 Alleyway in informal settlement, Buenos Aires 18
4.1 Woman wearing a *pollera*, Cochabamba 65
4.2 Parque Indoamericano, Sunday get-together, Buenos Aires 72
4.3 Parque Indoamericano, Sunday football matches, Buenos Aires 73
5.1 View from 50 m² flat where elderly carer worked, Spain 104
5.2 Daily elderly care work, taking the 'granny' for a walk, Spain 105
5.3 Three-storey house, Cochabamba 109
5.4 House being built with remittances, Cochabamba 110
5.5 Investment in a taxi, Cochabamba 111
6.1 *Locutorio*, internet café, accessing digital technologies to stay in
touch, Spain 125
7.1 Domestic violence campaign, Spain 148

Maps

1.1 Fieldwork sites 14
1.2 City of Buenos Aires and Greater Buenos Aires 15

Tables

3.1 Masculinity index (number of men to 100 women) and percentage
change for migrants from neighbouring countries living in
Argentina, by country of birth, 1980–2010   56
3.2 Migration destination, by sex   58
4.1 Social networks: Travel companion, by sex   68
4.2 Social networks: Accommodation upon arrival, by sex   69
4.3 Main reason for migration, by sex   84
5.1 Labour market participation, by sex, 2008   90
5.2 Current occupation: women   91
5.3 Current occupation: men   92
5.4 Occupation at destination, by sex   93
5.5 Money borrowed and saved, remittances, and income by
destination, US dollars   108
5.6 House deeds, by sex   108
6.1 Sexual division of labour, households by migration status   130

Acknowledgements

A project this long gathers many debts. This research started as a doctoral research project funded by the Centre for Development Studies, University of Wales, Swansea, which provided me with a maintenance grant within the scope of the Social Development Research Capacity-Building (SDRC) project, financed by the UK Department for International Development. Many of the initial fieldwork trips were undertaken in conjunction with work on this project, so I am very grateful for having had the opportunity to work on this project, alongside the late Peter Oakley, who was always very generous in providing advice and encouragement. I was then awarded a British Academy Post-doctoral Fellowship, which enabled me to revisit the initial fieldwork sites and extend the research to Spain. I am very grateful to the British Academy for offering their support at a time when research into South-South migration was still in its inception and longitudinal, multi-sited research was thin and far in-between.

In Cochabamba, the Centro de Estudios de la Realidad Economica y Social, CERES, was an initial point of contact and, while I was there, I benefitted from the support of Rosario León, Humberto Vargas, and Roberto Laserna. Later on, I collaborated with the Centro de Estudios Superiores Universitarios, CESU, at the University of San Simón, with María Esther Pozo, Alejandra Ramirez, and Leonardo de la Torre, while he was there, for which I am also grateful. Pablo Regalsky became a good friend during the course of this research; his warm hospitality and our shared interest in understanding change were invaluable. In Buenos Aires, I have enjoyed sharing initial research findings with Lucía Vera Groisman, Alejandro Grimson of the Universidad Nacional de San Martin, Sandra Gil Araujo at Gino Germani and later on, Jerónimo Montero Bressán, with whom I collaborated on a related project on migrants' involvement in grassroots organising. I would also like to thank Viviana and Adrian, for providing a 'home away from home', and the very friendly and helpful staff at the National Library.

Closer to my current home, I am grateful to my colleagues at the Global Development Institute for supporting my research in different ways and providing a friendly and collegial place to work in. I would like to thank Uma Kothari in particular, who mentored me during my postdoctoral fellowship.

Nina Glick Schiller and Dennis Rodgers have also offered friendship and a keen listening ear during critical points of what could have become a research saga, while Ron Skeldon, who examined my PhD, has now become a friend and co-editor of another large project that we are close to wrapping up.

I will never have enough words to thank those people who shared not only their experiences, but also their daily lives and homes with me. I am extremely grateful to Sahara Roque Rocabado and her family in Cochabamba and Buenos Aires, Marizol and David in Buenos Aires, Arminda and Miguel in San Fernando, and everybody else who welcomed me into their homes, took time out of their busy lives and did everything they could so that I was not only comfortable, but fed and safe while doing fieldwork away from my own family. My debt of gratitude to you is enormous.

My mum often helped with childcare while I was away on fieldwork and always believed that I would be able to achieve whatever I was trying to do, for which I am also grateful. Adair accompanied me during one of my field-work trips for my doctoral research. He continues to teach me how to better live my life by showing me what is really important. Islay came just in time to see this research project reach its conclusion and will, no doubt, help inspire future life ambitions. Juan has supported me in more ways than I realise, for which I will always be grateful.

Glossary of Spanish terms

Abanderado – A* student, adjective
adobe – bricks made of sun-baked mud
barrio – neighbourhood
bolsa, bolsa de viaje – literally 'bag' or 'travel bag', used to refer to the money that Bolivian migrants had to show migration officers to cross the border to Argentina
campesino – peasant
chau – bye
chorizo – sausage, Argentina
colectivo trucho – informal bus, Buenos Aires
comadre – fictive kin relation, female
Comibol – *Corporación Minera de Bolivia*, Bolivian state mining company
compadrazgo – practice of establishing fictive kin relations
compadre – fictive kin relation, male
compañera – partner, female
convivencia – cohabitation
cooperativista – member of the cooperative
corralito – restrictions put on currency withdrawals following the 2001 Argentinian crisis
cuadrilla – squad, in this case, squad of miners
empanadas – traditional Bolivian pasties, usually including chicken, egg, potatoes, sometimes olives
enrollado – traditional preparation of pork, Bolivia
escabeche – traditional preparation of pork, Bolivia
externa – domestic worker who is not a 'live-in' domestic worker
feminicidios – femicide, the killing of women because of their gender
ferias – street markets
ficha – card, in this case, cooperative membership card
fricasé – traditional preparation of pork, Bolivia
gente – people
hijita – daughter, diminutive

hombres mantenidos – literally 'supported men', used to describe men who are financially supported by their wives/partners

indígena – indigenous, indigenous person

indio – 'Indian', often used pejoratively

interna – live-in domestic worker

ley de migraciones – migration legislation

locutorio – internet café, phone booth

losa – flat cement roof, used in preparation for expanding an extra floor above, or floor tile

machismo – sexism

machista – sexist

madres desnaturalizadas – denaturalised mothers

malcriados – misbehaved

malo – bad

mamá – mum

mamá grande – literally 'big mum', grandmother

manda – from 'mandar', to order

media paga – literally 'half wage', extra wage given to employees twice a year, Spain

miradora – childminder, Spain

mita – compulsory labour recruitment system operating during the colonial period

monta – mounts

paisanos – co-nationals, Bolivians

palliri – women miners working outside the mine, usually breaking up ore for further processing

peón – labourer

pollera – traditional skirt worn in Bolivia

por el otro lado – the other way

por río – across the river

pulpería – local shop, mainly for Comibol workers, where they could buy all kinds of goods with a 'ticket'

villas, villas miseria – informal settlement, Argentina

zona roja – red zone, meaning dangerous area

1 Introduction

Feminist geographies

Feminist geographies of migration suggest that migration brings about social change, potentially disrupting patriarchal structures and bringing about new spaces where gender relations can be renegotiated and reconfigured. However, traditionally most studies have focused on changes that take place at destination, predominantly in the Global North. More recently, with the rise of interest in the so-called migration-development nexus, there has been greater attention paid to places of origin. However, there continues to be a bias towards studying migration from the Global South to the Global North. This book adopts a transnational lens and combines an analysis of regional, South-South migration with the more traditional South-North migration, undertaken by the same group of migrants. My aim is to analyse how gender, class, and ethnicity are renegotiated through internal and cross-border migrations. I do this through multi-sited itinerant ethnography conducted in two continents, three countries and five cities, and effectively across space with Bolivian migrants from the same community of origin. The analytical approach encompasses a multi-scalar and multi-sited intersectional approach to the study of social change through migration with the aim of investigating whether labour migration provides avenues for greater gender equality. In what ways do gender relations change through migration? What form does this change take? Can we define it as 'emancipatory'? And if so, for whom?

Bolivia has the highest poverty indices in Latin America and is increasingly dependent on remittances, which represented 5.4 per cent of Gross National Product (GNP) and 240 per cent of its Foreign Direct Investment (FDI). Furthermore, over a fifth of its population lives abroad (Banco Central de Bolivia 2007). Bolivia has received attention in relation to the socialist and pro-indigenous policies enacted by its first indigenous president, but most studies on mobility within and across its national borders are only available to Spanish speakers (Hinojosa 2008a, 2008b; Hinojosa 2009a, 2009b; de la Torre Avila 2006; de la Torre Avila and Alfaro Aramayo 2007; Román 2009; Solé et al. 2014; Cortes 2004) – for some exceptions see (Ryburn 2016, 2018; Rockefeller 2010). Yet the history of Bolivian geographic mobility represents a fascinating case study for better understanding the global process of feminisation of migration and the consequences this has for migrants and their communities of origin.

The feminisation of international migration flows has been linked to globalisation, the greater demand for women migrant workers, the ageing population in Europe and the US, and increased native women's labour market participation. Latin America is the first continent to achieve gender parity in its cross-border migration flows (Cortés Cartellanos 2005). Bolivian migration has been an exception to this regional trend, given that men have led its historic cross-border migrations. However, here too, there have been some significant changes and Bolivian migration has experienced a gradual process of feminisation since the 1980s. This book documents this change in a detailed analysis of life stories by both migrant men and women. The qualitative data is complemented by community surveys and census data to better understand how people's everyday choices and changed perceptions of mobility contributed to this gradual change as well as to how this change was lived at the personal level. The guiding questions therefore include the following: in what ways do gender relations change through migration? What form does this change take? Can we define this change as 'emancipatory'? And if so, for whom? Do changes associated with migration, such as greater labour market participation by women and restructuring of childcare through transnational social fields, lead to a fundamental change in feminine and masculine identities? Moreover, does migration lead to a structural change in gender relations and broader social transformation?

The methodology included multi-sited semi-participatory itinerant ethnography strongly influenced by feminist methodologies, global ethnography, and the extended case method (Burawoy 2000). Since 2000, I have visited Cochabamba ten times, Buenos Aires five times, and cities in Spain twice (Madrid, Algeciras, and Cadiz, see Map 1.1). During these visits, I recorded over one hundred interviews, mostly life testimonies with migrants and returnees, but also informal interviews with some family members, teachers, and community leaders. In addition, I also carried out two statistically representative neighbourhood surveys in 2002 and 2008 to gather quantitative data on socio-economic indicators and migration patterns. During fieldwork, I lived with key informants and shared their daily tasks. My analysis is therefore also based on extensive fieldwork notes.

Drawing on personal life stories to address the main questions identified above, the book contributes three main findings to our understanding of the relationship between gender, power, and space. First, the book shows how a focus on gender alone is insufficient. The book makes the case for the use of intersectionality given the interrelated nature of gender, class, and ethnicity. While this has already been argued elsewhere in feminist theory, the book shows with empirical data how gender, class, and ethnicity need to be considered within the same framework in order to understand the outcomes of processes of social change. In fact, it argues that women trade 'gender gains' for upward social mobility, a strategy that can only be understood by using an intersectional approach.

Second, by analysing change across space and time, the book also contributes to our understanding of how unequal social structures, such as the patriarchal family, are reproduced, albeit in a different form, through the migration strategies adopted by those interviewed. In this sense, the book contributes to a key question in the feminist geography literature – that of whether migration brings about opportunities for greater gender equality. However, it also makes a wider contribution to our understanding of how and why unequal power structures persist.

Third, the research also pushes the methodological barrier further by showing how longitudinal and multi-sited research brings about new understandings and new perspectives on issues, which have long been of interest, such as international migration. The methodology employed challenges the methodological nationalism prevalent in much migration research. This study highlights the interconnections of internal and cross-border migration, but also shows how those interviewed began considering cross-border migration because of their 'failed' internal migration and the unavailability of decent jobs in local labour markets. As a result, migrants invest the financial and symbolic resources they have accumulated through migration in consolidating their hold on the city, to finally belong to the city and live 'like people'.

While I have addressed some of these questions in single articles, this book opens up the opportunity to give greater prominence to migrants' life stories. At the same time, it also allows greater space to write the history of this neighbourhood and thereby analyse how this transnational community was created. The research project traced the creation of a transnational community from a mining town to Cochabamba, through internal migration and then to various regional and global migration destinations. My interest is in understanding how social inequalities, particularly gender and ethnicity, are challenged, created, and reproduced through the process of migration.

How crises created a transnational community

The barrio, the peri-urban neighbourhood where I began this research, used to be a milk farm located some nine kilometres from Cochabamba city centre. In the 1980s, a cooperative of miners bought this farm with the idea of using it to supplement the poor diets they had in the mining town. Malnutrition was rife, especially among children. Being situated over 3,000 metres above sea level, the mining town was too high for the production of vegetables or growing fruit trees. The intention was to improve access to a cheap source of milk from the valleys to improve the diet of the cooperative members and their families. However, during the mid-1980s, Bolivia faced an increasing number of challenges that led to the IMF-imposed Structural Adjustment Programme. The price of tin plummeted in international markets. Miners were one of the groups that were worst affected by the combination of the Structural Adjustment Programmes (SAPs) and the related privatisation of the state-owned mining company, Comibol, and lower mineral prices.

Many lost their jobs and could not continue making a livelihood in the mining town. They therefore converted the milk farm into individual plots and these were given to cooperative members to help them make the transition to the city. Cochabamba was at the time the third largest city in Bolivia and the miners thought that they would be able to make a livelihood by moving to the city.

This is how the barrio came to be, through internal migration from a mining town. Cooperative members who had made enough contributions to the cooperative, in terms of selling the mineral they mined through the cooperative, received an individual plot and began moving to Cochabamba. This is where I first came to know of this community, in the early 1990s. I was working as a volunteer with a religious community that had a house and a school right next to where the old milk farm used to be. At the time, the neighbourhood still looked like a farm and consisted of large fields, eucalyptus trees and only a scattering of adobe houses on the small 150 square metre plots. There was no access to electricity or running water and there were no roads, but there was a very strong sense of community.

I returned to the neighbourhood in 2000 when I was starting my doctoral research. As I was looking for my friend, trying to locate her house, two things became immediately clear. First, the neighbourhood had changed substantially in a very short period. Many more people had relocated to the neighbourhood. They had built new houses. However, people still knew where everyone lived. Second, many of the people I had known in the early 1990s were now in Argentina. My doctoral research project grew out of these observations and reignited my long-held interest in migration.

From 2000 onward, I returned to this neighbourhood on a regular basis and began tracing some of its people in Buenos Aires and later on in Spain. As my research project grew and diversified to include a generational perspective, so did my involvement with the community. Some ten years later, when I started participating in the annual dance for the neighbourhood saint's festivities around the 1st May celebrations (see Image 1.1), the neighbourhood was unrecognisable from the one I had first known in the early 1990s. Most houses were made of brick, often multi-stories, with tiled roofs, indoor kitchens and toilets, running water and electricity (see Image 1.2). Another significant change was that many people did not necessarily know their neighbours. People who had come from the mining town still recognised each other, but other residents had moved to this neighbourhood. Was this still a 'community'?

In the interviews and the informal conversations I held with residents, it was clear that the identity of being somebody from the mining town was still strong, even among those who had only been born there and then grew up in the city. Indeed, feelings related to their identity as miners are still strong and many supported the mobilisations of cooperative miners organised by FENCOMIN in July and August 2016. However, I had some doubts about whether we could still talk about the neighbourhood as a community.

Image 1.1 Dancing the *diablada* at the annual neighbourhood festivity, Cochabamba

This move to Cochabamba was, however, fraught with problems, from high levels of unemployment to the stereotyping of miners as dangerous, violent drunks. At one point during fieldwork, for example, somebody living in the city centre challenged me for living in a *zona roja* or red zone, a neighbourhood perceived as dangerous by those who do not live in it. However, I have never felt exposed or in danger, except for the dogs that roam around the neighbourhood after dusk. Failing to secure their livelihoods in Cochabamba meant that many started looking for work elsewhere, particularly in Buenos Aires, but also the Chapare and Santa Cruz (in Bolivia), Israel, or Guatemala. Buenos Aires was an attractive destination during the 1990s because of the convertibility plan, which pegged the Argentinian peso to the US dollar. However, Bolivians, Paraguayans and Peruvians in Buenos Aires, the capital of Argentina, experienced high levels of racism and xenophobia (Grimson 1999; Hinojosa forthcoming; Bastia and vom Hau 2014). National public discourse blamed migrants from neighbouring countries for high levels of unemployment, crime, and a cholera epidemic, in an attempt to exonerate the public authorities from their responsibilities for solving structural problems in the economy and political culture. When the 2001 crisis struck, migration streams reversed momentarily and diversified again with newer opportunities opening up in Spain and Italy, for example, but Argentina remained an attraction pole for regional migrations. Following the financial downturn in Europe, many Bolivian migrants returned to Bolivia or to their historical

Image 1.2 View of the neighbourhood, Cochabamba

destination, Argentina, but other opportunities also opened up in Brazil and Chile, diversifying regional migration streams (Baby-Collin and Cortes 2014).

When looked at from the Bolivian or a Southern perspective, crises are not exceptional. They occur with regularity. They bring additional suffering, the need to readjust, and in this process migration is instrumental, for those who can afford it, in helping people diversify their livelihoods and increase the chances of survival. Given the many resources needed to undertake migration, 'survival' does not necessarily refer to just physical survival but the maintenance of a way of being, which might also include projected upward social mobility. For many of the migrants I spoke to, migration was less about physical survival and more about their wish to achieve some status and dignity through their migration projects. In particular, they wanted to be accepted as *gente,* or people. This word has strong urban and class connotations, more closely associated with what we might term 'urban citizens' (Tapias and Escandell 2011). The wish to be accepted as people can be understood to be a reaction to the stigmatisation my interviewees experienced as miners when they first arrived in Cochabamba.

Gender and itinerant migrations

Amid this general picture of change in the neighbourhood, I was particularly interested in gender relations. From an initial question of how gender relations influence migration flows, the research project evolved to address changes in gender relations brought about by migration. This is clearly a circular question experiencing a constant process of change as well as multiple causes. Some changes may be brought about by migration while others might be the result of wider socio-economic change, lobbying by women's movements, and greater awareness of gender inequality as a result of participating in local women's grassroots movements that are not necessarily linked to migration. This relationship is further complicated given that migration itself generally results from wider processes of socio-economic change. As is clear from the above, internal migration from the mining town to the city was caused by wider changes in socio-economic and political structures, such as the Structural Adjustment Programmes, changing prices in minerals, and the decreasing importance that mining played in the restructured Bolivian economy. Yet it is possible to identify specific changes that are brought about through migration. Pratt and Yeoh (2003), for example, talk about migration potentially opening up new spaces for the creation of more equal gender relations:

> there tends to be both a deep utopic hope that transnationalism may offer opportunities to realign and equalise gender relations, and a knowing scepticism that patriarchal relations return in different guises in different times and places. (161)

In the surveys as well as the life stories, therefore, the focus became to try to identify such spaces where migrants as well as non-migrants could create more equal gender relations. This was an exercise of exploring the ideas, dreams, and possibilities envisaged by the migrants I interviewed in terms of how they see women and men relating to one another as well as new possibilities for being a man and a woman. How have these ideas emerged? Where did they take their inspiration? What kinds of changes have they experienced in their lives that then gave rise to new possibilities?

These questions were embedded in the new experience of Bolivian migration, especially the greater participation of women in international migration. While Argentina was the historic main destination for Bolivian migrants, this all changed radically with the 2001 Argentinian crisis. With the peso no longer pegged to the US dollar, Bolivian migrants started looking for new destinations and shifted their migration routes towards Spain, a process that opened up opportunities for multi-polar migrations (Hinojosa 2008b). This shift was accompanied by a strong feminisation of Bolivian migration flows (Hinojosa 2008a). While men had historically dominated migration towards Argentina, women led the new migration towards Spain. In 2008, women represented 55.9 per cent of Bolivian residents in Spain (INE 2010). This shift gave rise to moral panic in the local media and conservative quarters such as the Catholic Church, which identified women's greater participation in international migration with the dissolution of the traditional (patriarchal) family and raised concerns for the well-being of the migrants' children. At the same time, from more progressive quarters, women's migration was seen as a sign of empowerment with women seeking their freedom and autonomy through migration. From this point of view, women's migration was a sign of emancipation. It represented women's opportunity to liberate themselves from patriarchal oppression, whether this was related to personal relationships or work opportunities. There is certainly evidence to show that gender roles changed, sometimes significantly, because of migration. However, it is also timely to ask, fifteen years on, whether these processes also brought about greater and longer-lasting changes in gender relations.

This book provides an original contribution to the migration literature by addressing changes in gender relations and migration transnationally, by following a group of migrants from origin to various destinations. In particular, it responds to recent calls for expanding our understanding of migration by starting to consider migration processes other than the well-studied movements from the Global South to the Global North (Castles 2010; Kofman and Raghuram 2012, 2015; McIlwaine and Ryburn 2018). Multi-sited and extended fieldwork in Cochabamba, Buenos Aires, and three cities in Spain provide an innovative and different perspective on the changes in gender relations brought about by migration. First, because I reverse the usual focus on countries of destination and countries of the Global North by focusing on countries in the Global South. Second, because I complement the study of migration from the Global South to the Global North with a focus that combines regional and global migration flows. Third, because the migration chains are followed through from origin to destination and back to origin, thereby also integrating an understanding of the process of return and the ways in which changes brought about by migration are incorporated, or otherwise, in migrants' everyday lives following their return to their country of origin.

With this in mind, there is also a note of caution. To make sense of this diversity and fluidity in migration flows it was also necessary, to some extent, to simplify and stylise movements within and across international borders. Migration is seldom as clear-cut as I describe above: regional international migration follows internal migration, then both of these turn into global migration (South-North) as a result of the crisis and the cycle completes with a return. There are migrants who have not returned, for example. Others 'skipped' the internal migration part and migrated directly from the mining town to Buenos Aires. Some stayed in Buenos Aires; others stayed in Spain. Still others did not go to Spain and stayed in Bolivia following their migration to Buenos Aires. Others went to Chile or Brazil after or instead of Argentina. Some went to the United States, Guatemala, Israel, and Russia. Yet we lack a language to make sense of this diversity. So for the sake of analysing what goes on with gender relations, we simplify migration flows and represent it as following some logical and progressive steps: from internal to cross-border (regional), then global international migration and return.

The life stories that are included in this book will certainly speak to the greater diversity and complexity included in migration processes and I will highlight this wherever possible, but I do refer to these different types of migration to try to distinguish the different migration phases that the transnational community has gone through. While individual people may have followed very different migration patterns, the community itself could be said to have gone through these different stages: internal, regional, global, and return.

Migrants' experiences clearly indicate that migration, even when migrants return, is seldom if ever a completed project (de la Torre Avila 2006). Migrants keep their options open and migration is always on the horizon as a possibility when times get tough again, when a child's illness leads to unexpected large hospital bills, or when it becomes impossible to pay one's debts with their Bolivian salary. Migration was part of the way the Spanish colonial rule organised its Andean colonies, with indigenous people regularly spending parts of the year in domestic service (women) or in mines (men) through the *mita* system. Andean indigenous communities continue to aim to tap into different ecological zones to increase their access to different growing climates while at the same time decrease risks to ecological disasters. Contemporary internal and international migrations represent a very similar way of organising communities and household economies with the aim of increasing opportunities by tapping into distant labour markets while at the same time decreasing risks. Migration therefore becomes, as it has always been, a way of life, a necessary journey, an itinerant migration – travelling from place to place, but always maintaining a tie with the place of origin.

Bolivian migrations in context

Clearly, migration is not a new phenomenon. Estimates suggest that about a fifth of the Bolivian population resides abroad, with the largest number in Argentina (Jones and de la Torre 2011; Hinojosa forthcoming). Current migration flows are rooted in much older, historic migrations, which have been practiced for centuries by Andean indigenous communities. However, with globalisation, they have taken new forms, expanded in scope and taken on new meanings. Yet the logic behind the movement of people and the spreading of the household across different ecological (then) and economic (now) levels has remained the same (Hinojosa 2009b).

Bolivian cross-border migrations went from being male-led, or male-predominant to the most recent female-predominant migrations to Spain (Parella 2011). Although the term 'male-led' has been used in the literature to refer to the greater proportion of men in a specific migration flow, the term more correctly refers to who leads a migration stream and who takes the main decisions. The term 'male or female-predominant' refers to the numerical proportion of men or women in a migration stream (Donato and Gabaccia 2016). Taking this distinction into account, the newer migration to Spain for Bolivians can be said to have been both 'female-led' and 'female-predominant'.

For most of the twentieth century Bolivian migration was of a typical cross-border characteristic. Bolivians migrated from the south of Bolivia to the north of Argentina, predominantly from rural areas, to work on agricultural plantations. Rural communities, lacking sufficient land and water to make a living, sought seasonal work on sugar and tobacco plantations. This started changing with the mechanisation of production, which decreased the demand for labour, which went hand in hand with an expansion of demand for service provision in the capital, Buenos Aires, and other destinations within Argentina later on (Sassone 1989). What, for decades, was a typical and quite constrained cross-border migration, therefore began shifting its destination towards the capital of Argentina (Cortes 2011). Bolivians inserted themselves in domestic work and construction, as well as manufacturing sectors, particularly garment workshops (Magliano 2007; Bastia 2007). Bolivian migrants also started working their way up the horticultural production sector that supplies Buenos Aires (Benencia 1997).

During this time Bolivia was also experiencing deep political and economic changes, which were implemented because of the Structural Adjustment Programme imposed by the International Monetary Fund. While the vast majority of migratory movements were confined, until then, to the southern border area, this started changing with the impoverishment and 'relocalisation' brought about by the economic restructuring. A greater proportion of the Bolivian population started participating in regional migration flows, particularly from the departments of Cochabamba and La Paz. Migration, therefore, went from being a cross-border affair to one in which the whole nation participated.

This is not to say that these were the only migrations that took place in Bolivia. With its rich history in mining, specific migration streams also developed around the mining industry. Miners used to migrate from the mining towns in the department of Oruro and Potosí to find work in Chilean mines (de la Torre Avila 2006). There were other, more skilled migrations to the US (Grimson and Paz Soldan 2000). A Bolivian community exists in Washington, for example, which has over time diversified to include Bolivians with a rural background (Whitesell 2008; Strunk 2014; Yarnall and Price 2010). However, the vast majority of its migrants opted for regional migrations, particularly to Argentina, given that this was the most affordable and accessible stream.

Migration to Argentina has been marked by informality, discrimination, and xenophobia. Despite the fact that Argentina returned to democratic rule in 1983, it took over two decades to change the migration policy that had been imposed during military rule. During this time, migrants from neighbouring countries continued to migrate to Argentina, but enjoyed few rights. During the 1990s, the authorities blamed them for the increasing crime rates, unemployment, and health scares that afflicted the country. While all migrants from neighbouring countries suffered the consequences of discriminatory and xenophobic discourses, Bolivians were more easily targeted because of their indigenous ancestry. The xenophobic wave was then compounded by the racialisation of the Bolivian community, particularly in Buenos Aires (Grimson 1999; Bastia 2015b; Hinojosa forthcoming). This period culminated with the 2001 Argentinian crisis, during which migrants from neighbouring countries but also Koreans, were targeted and their businesses ransacked.

Civil society organisation had during this time been active in resisting the scapegoating of the migrant population. In the early 2000s, they were successful in putting forward a proposal for a new migration policy that would give migrants the right to live and work in Argentina. The government approved the *Ley de Migraciones 25871* in 2003 and the new legislation came into force in 2004, giving migrants full access to social and civic rights, including protection against automatic deportation. The new migration legislation provided a legal framework for the protection of migrants' rights, including the opportunity to legalise their stay in Argentina. It promoted the principle of non-discrimination, guaranteed migrants and their children access to education and social services, independent of their legal status. The new migration legislation was, at the time of its implementation, deemed a true success not only regionally, but also globally, given the general trend of closing national borders.

The new migration legislation was followed by a regularisation programme, called Patria Grande, implemented in 2006, aimed at regularising the stay of migrants from neighbouring countries. Almost half a million people (423,697), mostly from Paraguay (58 per cent), but followed by Bolivians (24 per cent), applied for the regularisation programme by 2010 (Ministerio del Interior and DNM 2010).

Following the Argentinian crisis in 2001, there was a marked re-organisation of Bolivian migrations. The Bolivian migration to Spain increased rapidly from just a couple of thousand people in 2000 to a peak of 242,496 in 2008, followed by a steady decline to 102,550 in 2017 (INE 2017). At the peak in 2008, women represented 55.9 per cent of the total Bolivian population, but this increased to 58 per cent for the years 2011–2014, indicating a greater return of Bolivian men following the financial crisis given their higher likelihood of becoming unemployed (INE 2017; Parella et al. 2014).

Spain was at the time experiencing high rates of economic growth with a relatively open border to new migrants from Latin America. People travelled as tourists. All that was required was a passport, a hotel booking and some money to show that they had sufficient funds for their 'holiday'. They obtained their tourist visa at the border. They then overstayed their visa. Bolivians, particularly women, took the opportunity and started migrating to Spain in large numbers. They found work in the domestic and care sectors, mostly working as domiciliary elderly carers. Bolivian men also migrated to Spain but in smaller numbers. Given that they had entered as tourists, they

had no work permit as such, so migrants undertook most of the work informally. The relatively easy access to the Spanish labour market changed in
April 2007 with the introduction of a visa for Bolivians, which migrants had
to secure before travelling to Spain. By then, the economic situation had
deteriorated and most of the incentives for migrating to Spain vanished with
the financial crisis that followed (Bastia 2011). The timing of my fieldwork,
during the summer of 2009, therefore coincided with the financial crisis and
the worsening of the economic situation in Spain.

Intersectionality

Despite its focus on gender relations and its relationship to place, the book
adopts intersectionality as an invitation to integrate gender, ethnicity, and
class. When I began this research it soon became clear that the initial focus on
gender alone was insufficient to explain migrants' experiences in Argentina.
The gender relations that influenced migration flows were mediated by ethnicity and class. Migrants used their own social networks to find out about
opportunities in Buenos Aires. I wanted to explain how these social networks
were gendered, that is, how men and women had different access to social
networks, but I could not do that without explaining the role that ethnicity
and class played in building social networks. Moreover, the dominating features of the life stories recorded in the early 2000s in both Cochabamba and
Buenos Aires related to ethnicity and race-based discrimination. There was
little mention of gender. Given my interest in trying to convey the story as it
was told, it became impossible to leave out ethnicity and the discrimination
that Bolivians experienced in Argentina during the 1990s, as this permeated
their everyday lives and the way in which they made sense of Argentina.
While I initially merely added race, ethnicity, and class to my conceptual
framework, I later started using intersectionality as a way of integrating the
gender analysis with these other dimensions of difference.

Intersectionality emerged from black women's movements in the US during
the 1980s as a critique of the white women's feminist movements. It was first
proposed by Crenshaw (1991) and was then taken up by other critical race
theorists, to counter the essentialising notion that gender is the most important oppression experienced by women. Those who adopt intersectionality
reject the notion that gender, race, and class are essentially distinct forms of
oppression. They argue that these forms of oppression intersect and propose
intersectionality as a way of bringing light to intra-group differences that were
obscured by essentialising approaches.

Intersectionality has become particularly popular in migration studies since
the turn of the century, when European feminist scholars began adopting
intersectionality as an approach. While migration scholars interested in
understanding how gender relations influence migration flows have made
advancements in this field, the focus tended to remain on women as subjects
of migration or on gender, with little understanding of how gender is also a

classed or ethnicised concept. An increasing number of studies are now adopting intersectionality while addressing migration in the European context (Lutz et al. 2011; Ludvig 2006; Prins 2006; Burman 2003; Buitelaar 2006; Kosnik 2011). These new approaches address some of the shortcoming of previous studies of gender and migration. For example, they highlight intra-group differences and are able to show how different groups of women adopt contrasting approaches to challenging gender regimes (McIlwaine and Bermudez 2011). Others are able to focus on privilege, to complement the usual focus on disadvantage (Kynsilehto 2011; Riaño 2011). Intersectionality is therefore able to provide new perspectives on our understanding of the relationship between gender and migration. However, many studies are able to achieve the same but without the explicit adoption of intersectionality as an approach. Some working from a post-structuralist perspective, such as Pratt (1999), have been able to unpack classed and racialised notions of gender but without adopting intersectionality as an explicit framework of analysis (Pratt 1999). It is with this in mind that this study adopts intersectionality as an invitation to address the classed and racialised notions of gender, but by integrating historically informed, local understanding of gender, ethnicity, and class.

Methodology

The methodology was multi-sited and longitudinal. It consisted of in-depth life story interviews with migrants and returnees carried out in Cochabamba, Buenos Aires, and three cities in Spain (Madrid, Algeciras, and San Fernando-Cadiz) over a period of ten years (see Map 1.1). During this time, I recorded over 100 interviews with both men and women. These were complemented with two surveys of the peri-urban neighbourhood in Cochabamba carried out in 2002 and 2008. The survey applied a household questionnaire to every third household in the neighbourhood to gather information about basic socio-economic characteristics of the neighbourhood as well as migration trends. The 2002 survey included 157 households and the 2008 survey 171 households. I carried out all the interviews and transcribed most of them. The names included in this book are all pseudonyms. I have also changed some details so that interviewees could not be identified.

I benefitted from the help of a good friend of mine for the questionnaire on both occasions. The questionnaire included both closed and open questions. I also carried out informal interviews with community leaders, teachers, NGO workers, academics, and members of a religious community that live next to this neighbourhood.

Building trust with potential interviewees was essential, especially in Buenos Aires, where Bolivian migrants were quite defensive and diffident. Here, again, my friend played an essential role in introducing me to Buenos Aires and the places where Bolivians live and work. She had lived in Buenos Aires for two years during the 1990s. In 2002, we travelled together to Buenos Aires where

Map 1.1 Fieldwork sites, map by Cartographic Unit, University of Manchester

we stayed for a week and then again in 2003, when she stayed for a month. She introduced me to the places where she had lived and worked. These were informal settlements, the *villas miseria* or *villas* as they are known locally: Villa 20 – Lugano, Parque Indoamericano, Liniers, and the informal street markets in Cildañez, Soldati and Villa 1–11–14 in the Southern neighbourhoods of the City of Buenos Aires; as well as Villa Celina, Burje, La Salada, and La Ferrere in Greater Buenos Aires.

The first time I travelled to Buenos Aires, I undertook the same three-day journey by bus that most of my interviewees had undertaken. As a result, I got to know Buenos Aires through migrants' eyes: sharing their first journey and impressions of Buenos Aires and the *villas* where many Bolivians and other migrants from neighbouring countries live (see Images 1.3, 1.4, and 1.5). Here as well as in Cochabamba I lived with members of this community, which allowed me to get to know other people and participate in their daily activities. Going to the market, hanging out at home, walking children to

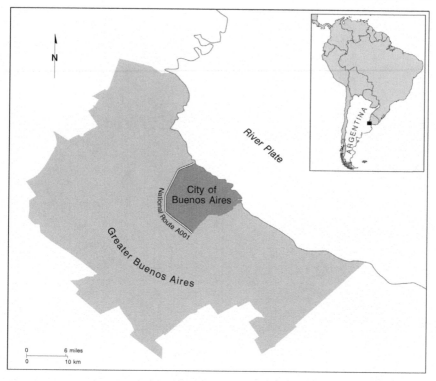

Map 1.2 City of Buenos Aires and Greater Buenos Aires, map by Cartographic Unit, University of Manchester

school or going to funerals, played an important part in gaining the trust of potential interviewees as well as a deeper understanding of the context in which these migrants lived. Life story interviews were already quite rich but without living and sharing at least part of the daily activities of the migrants' lives would have led to a much narrower and more limited understanding of the stories and experiences they shared with me.

After a month, I stayed on for another month, on my own, but others took on the same role that my friend had played. While living in informal settlements presented its own set of challenges and risks, the people I stayed with always made sure that these were minimised. They told me where to walk and which alleyways I should avoid. They pulled me off buses when they felt I was in danger of being mugged. They walked me to interview places and picked me up from bus stops so that I would not have to walk into an informal settlement on my own. We worked out strategies to keep the interview tapes safe, when digital recorders were not yet available, such as taking a child with me to the interview and hiding the tape in his socks or nappies and replacing the interview tape with a blank one. In case I got mugged, I would lose the recorder but not the interview. In any case, I was much less likely to be hassled when I was accompanied with somebody carrying a baby or

Image 1.3 Informal settlement, City of Buenos Aires

when walking with a child. My interviewees also fed me and provided me with sleeping space. One time they managed to do so by finding a space for me in a room that was already used by a whole family consisting of four children and two adults. In view of everything they gave me, I feel I gave very little in return: mainly a keen listening ear and somebody to share their experiences with. Sometimes interviewees asked me to accompany them to hospitals or offices, particularly in Buenos Aires. They felt that somebody who was European would be able to get them better attention and fairer service. This book is a debt I owe many of my interviewees who were keen to see the story of their community told in printed form.

Since 2013, I have been involved in two research projects that have wider objectives, but have also informed part of the analysis included in this book. The first one, carried out with Jerónimo Montero Bressán, focused on the role that migrants played in grassroots organising in Buenos Aires' informal settlements. We carried out 18 interviews with migrant and grassroots organisations. These are not reported directly, but the outcome of that project informs part of the analysis included in the section on domestic violence (Chapter Seven). The second one, initially carried out with Maria Esther Pozo of the University of San Simon, Cochabamba, seeks to understand the consequences that migration has for the migrants' parents. As part of this project I carried out a small number of interviews in the *barrio* and these have informed the analysis included in sections on care and social reproduction (Chapter Six).

Image 1.4 One of the fieldwork sites, City of Buenos Aires

Structure of the book

The book is structured around four key themes, which recurrently emerge in the gender and migration literature: mobility and social networks, work, care, and intimacy.

Chapter Two traces the evolution of thinking around gender and migration, from the initial feminist critiques of mainstream migration theories and attempts at highlighting the fact that women are also migrants, to the current interest in intersectionality and the ways in which gender intersects with race, class, and ethnicity to produce unequal outcomes for women migrants.

Chapter Three provides some empirical context to the transnational community under consideration in this book. It describes the mining town where many of the interviewees were born, or where they trace their origins, the neighbourhood on the outskirts of Cochabamba as background to the gendered mobilities that are considered in the chapters that follow. This chapter shows how 'the transnational community' was created through multiple crises and how migration was a direct response to the economic and political crises in Bolivia, Argentina, and Spain.

Chapter Four focuses more directly on gendered mobilities and social networks in particular. It shows how mobility, including border crossings, are gendered and goes on to consider social networks and the different ways in which men and women access and use social networks through their migration

Image 1.5 Alleyway in informal settlement, Buenos Aires

journeys. Much of the early literature on gender and migration has given a lot of importance to the sequencing of migration – or who goes first – as a proxy question around who decides about migration. In this chapter I therefore consider sequencing and decision-making as a way to dispel the myth that sequencing in some ways is representative of the decision-making processes in migration. Here I show that male-led migration journeys are not necessarily the outcome of men's greater decision-making power and that some women-led journeys are often decided for by others, not the woman undertaking the migration journey. Sequencing, therefore, is not necessarily directly linked to autonomy in the decision-making of migration journeys.

Chapter Five deals with the world of work. Given that this is predominantly a labour migration, in this chapter I explore the types of paid work that women and men migrants engage in. How do those gendered social networks explored in the previous chapter mediate the types of jobs available to men and women? How are the Argentinian and Spanish labour markets gendered? And where are women and men migrant workers from this community able to find work? I describe the types of jobs that they engage in and show that, while both men and women often find work that entail exploitative working conditions, women have a more limited choice to make with regards to work. However, these gendered jobs also pave the way for a potential renegotiation of gendered relations among members of this transnational community.

This engagement in transnational searches for work and the spreading of livelihood strategies across national boundaries also entail a renegotiation of social reproduction. Chapter Six therefore focuses on the world of care and how migrants negotiate social reproduction through the migration process. Who cares for whom in this context? How does this differ from traditional ways of organising care among Bolivian families in similar peri-urban areas? Do these changes entail more profound renegotiations of gendered identities? While answering these questions, this chapter also provides an extension to the global care chains framework, by including men, types of jobs other than just care work, and a historic analysis and one committed to understanding the organisation of care within specific socio-cultural contexts. In line with other studies that have recently emerged from the Latin American region, I conclude that a broader approach to social reproduction arrives at a less pessimistic view than the original global care chains framework.

Chapter Seven then turns to intimacy and how intimate relations change as a result of migration. Taking a broader understanding of intimacy, one that also includes relations between parents and children, and those that are forged in paid employment, this chapter shows that migrants' intimate relations are profoundly shaken by the experience of migration. This is not to say that migration always entails negative consequences for migrants' intimate lives. While some intimate relations are weakened, others are forged through the migration process while others still merely change shape and characteristics. Here I also pay attention to how unhappiness in intimate lives also gives rise to a desire to migrate, particularly for women who find themselves in violent relationships.

Chapter Eight ties these questions together in a broader analysis of social change and gender relations in this transnational community. Almost twenty years on since this research has started, what has changed in this community as a result of migration? Do those changes in gender relations evident through the migration process entail a deeper renegotiation of gender relations and gender identities? How does intersectionality help us understand these changes?

A note on the use of 'community'

Before moving on, I would like to return to the 'community of origin' and defend the use of the term 'community', despite its many drawbacks, if not only to acknowledge that through migration the community was initially reconstituted: migration helped the community in Cochabamba survive (see also Strunk 2014). However, migration also contributed to its un-making, given the large differences in returns from regional or global migration projects as well as increasing differentiation based on educational achievements and permanent out-migration by selling the original plots of land (Bastia 2013b). New internal migrants who move to this neighbourhood do not have the shared history of having lived and worked in the mining town and while the mining culture is still strong, the increasing differentiation within the neighbourhood is stretching the meaning of 'community'. During the 1990s, before Facebook and WhatsApp became widely used, the death of a young woman in a motorbike accident was widely shared within the neighbourhood so that the experience was lived simultaneously in Buenos Aires and Cochabamba. However, today's highly differentiated neighbourhood lacks the type of social networks that would allow such a story to 'travel' in the same way. While the focus remains on gender relations, and individual migrants, paying attention to the ways in which these migrants relate to the community has provided a rich basis on which to build my analysis and understand how they relate to their neighbours and to the multiple places to which they feel they belong.

2 Gender, migration, and social transformation

Feminist geographies of migration[1]

Feminist geography has always been interested in the interaction between social relations and place. Therefore, migration and the changes brought about by people's movement through space have always been high on the feminist geography agenda. In much of this work there is an implicit or sometimes an explicit hope that migration will bring about social change, potentially disrupting patriarchal structures and creating new spaces where gender relations can be renegotiated and reconfigured (Hondagneu-Sotelo 1994, 2000; Pessar 1999; Pessar and Mahler 2003; Pratt and Yeoh 2003; Pratt and Yeoh 2004b). As such, it raises direct questions in relation to the imbalance of power relations of gender, class, and ethnicity, including issues of justice and the distribution of resources in a given society.

Much of this literature has explored these changes in countries of destination and often from a liberal feminist perspective. Such a perspective frames gender relations in Western, high-income countries as inherently more equal than in the lower-income countries of the Global South. However, clearly, gender relations in countries in the Global North are also unequal, although these gender-based inequalities may take different forms. Gender relations do not only change as a result of migrants moving from the Global South to the Global North or from poorer to richer countries. They can also change when people gain greater awareness of the socially constructed nature of gender relations. Moreover, migration leads to changes in gender relations, these changes may not always be positive in the sense of leading to increased empowerment. Women may also experience a strengthening of gender inequalities as a result of migration (Gregorio Gil 1998).

However, asking questions about changes in gender relations as a result of migration need not be framed within a liberal feminist perspective. These questions, or at least the way I pose them here, relate more to a recognition that awareness of gender inequalities can be awakened by going to another place where gender relations are organised differently and take a different form. This could be another place within the same country, going to another place where the dominant ethnicity is different or, indeed, where gender relations are organised differently based on class. The ways in which migrants might relate to these differences will not be as dramatic if they were already aware of how gender relations were different as a result of, for example, class. Men from Bolivian indigenous communities, who grew up within a context where gender relations were complementary and women's roles were valued within the household economy, are exposed to quite different ideals of masculinity when they enter military service (Canessa 2012). However, this awareness of the different ways in which gender relations are organised, does not lead them to question the social construction of gender inequality. My contention is that with migration, especially transnational migration, this is more likely to be the case, because migrants are exposed to practices that they were less likely to have been familiar with or aware of before migration.

Migration often also leads to changes in gender roles. When men migrate, the spouse who is 'left behind' may gain greater autonomy in running the household production, managing finances or greater freedom of movement locally (Cortes 2004). However, gender roles are truly shaken when women migrate for work and trigger a reversal of gender roles. When women become breadwinners, they pose a direct challenge to men's traditional and expected role of providing financially for the family and this may also lead to women's greater empowerment (Parella 2011; Baby-Collin et al. 2008; Kabeer 2000).

Gender as necessary but not sufficient

Two emergent issues spurred a wave of interest in the role of women in migration during the 1970s: the realisation that women were migrating in ever-greater numbers and the fact that they displayed higher labour market participation rates when compared to their situation before migrating or those of the 'native' women (Pessar 1999). The second wave of the feminist movement at the time was contributing to a broader critique of social science in general. These feminist critiques were then successful in highlighting the very basic fact that women were also migrants.

In her review of the relationship between feminist and immigration scholarship, Hondagneu-Sotelo (2000) describes the gender and migration scholarship as being divided in three stages. In the first stage, which took place in the late 1970s and early 1980s, feminist migration scholars attempted to remedy the exclusion of women from migration studies by focusing on women migrants. This approach rested heavily on the sex role theory in which women's activities are seen as complementary and functional. As such, these early approaches focused on women but not on gender relations, despite the fact that the analysis did address power inequalities, for example, in accessing labour markets.

At this stage, the first step was towards constructing a more gender-balanced view of migration by including women in the analysis, in effect, highlighting the fact that women too are migrants by making them visible (Brettell and Simon 1986; Lutz 2010). Until the 1970s, it was assumed that men were the primary, economic migrants and that women generally migrated as secondary, or associational, migrants (Donato et al. 2006).

Typical contributions from this early stage were those included in Phizacklea's (1983) edited volume on the role of women in labour migration from Southern European and developing countries to Western Europe, linking women's status as migrants to racial and legal subordination. A number of key publications began presenting a critique of migration studies from a feminist perspective, highlighting women's role in migration (Anthias 1983; Morokvasic 1984; Phizacklea 1983).

During the early 1990s, it was still common for migration studies to work with the assumption that migrants are single men (Pedraza 1991). This is despite the fact that in migration to the US women have outnumbered men in legal migration as early as 1930. Women also 'lead' some migration streams such as those of the Turkish migration to Germany (Brettell and Simon 1986) or Peruvians to Spain (Ioe 1991).

Making women migrants visible was already an improvement, given that, until then, migration studies had been gender blind and effectively biased against women, despite Ravenstein's early acknowledgements of differences in how women and men engaged with migration, when he argued that:

> Woman is a greater migrant than man. This may surprise those who associate women with domestic life, but the figures of the census clearly prove it. Nor do women migrate merely from the rural districts into the towns in search of domestic service, for they migrate quite as frequently into certain manufacturing districts, and the workshop is a formidable rival of the kitchen and scullery. (Ravenstein 1885: 196)

Until the 1980s, the dominant discourse within migration studies dressed economic migrants as men, replicating long-held views of men's dominant and active role within households and labour markets, while assigning women passive roles of care-givers and 'followers'. Women were seen as passive migrants, following their partners to their destinations or staying in their places of origin. These prejudices and the omission of gender from migration research clearly also extended to migration policies, which were generally focused on the assumption that most labour migrants are men (Kofman 1999). However, even this early on, there was a recognition of diversity within migration studies, arguing that family reunion and female occupational structures are much more complex and diverse than is generally acknowledged:

> Women have ambitions and strategies that cannot be reduced to the simple division between economic and personal autonomy, on the one hand, and family migration, on the other. One strategy does not preclude other meanings, intentions and strategies. (Kofman 1999: 287)

Interest in gender and migration began through the analysis of structural elements that influenced the characteristics of particular migration flows. Gender-selective migration is the term that describes movements of people that are differentiated by gender (see Chant and Radcliffe 1992). It refers to the fact that women and men migrate in different numbers, to different destinations and for different lengths of time. It also refers to the fact that gender relations prevalent in the place of origin and destination shape the migration process in such a way that women and men have different experiences of their respective movements. Chant and Radcliffe (1992) identified three main reasons that explain regional difference in gender-selective migration patterns: (i) the relative involvement of women in the agricultural sector; (ii) the relative demand for female labour in the cities; (iii) the extent to which women's mobility is constrained by social and cultural construction of gender relations, informal institutions, and cultural practices.

Such categorisations are useful for identifying important factors in migration, but the relative importance of each determinant will vary according to the specific context within which migration takes place. Moreover, other issues, such as the presence of established social networks or long-term historical relations between different countries, will also influence who migrates and where to, as I discuss in detail in Chapter Four.

During the second stage, between the late 1980s and early 1990s, research focused on the role of gender and migration; specifically on the gendering of migration patterns and the effect of migration processes on gender relations (see also Willis and Yeoh 2000). Many studies during this period looked at intra-household power relations and decision-making as well as the gendered nature of social networks.

Attention later turned towards gender, both as a constitutive element of migration as well as in terms of how migration challenged existing gender relations (Donato et al. 2006; Silvey 2006; Curran et al. 2006). However, there was little consensus over the way gender relations shape migration movements. Some have suggested that the increase in the proportion of women migrants was a result of weakening patriarchal structures (Gugler 1997). Others have argued that it is precisely because of the existence of patriarchal structures that gender-selective migration takes place (Radcliffe 1993). It follows that, unless gender-selective migration creates opportunities for challenging patriarchal structures, these will be maintained, as will be discussed later on in this chapter. However, the relationship between migration and gender is both complex and analytically challenging (Pessar 1999).

By the turn of the century, we could count on a wide variety of studies undertaken from a gender point of view and which covered a whole range of issues. Important collections of these became available: on gender relations in developing countries' migration flows (Chant 1992); on migration and gender in developed countries (Boyle and Halfacree 1999), as well as more general overviews on gender and migration (Willis and Yeoh 2000).

Despite huge advancements both at the theoretical as well as the empirical level in terms of our understanding of migration as a gendered process, three problems persisted. First, mainstream migration studies continued to show resistance to the uptake of the feminist critique and remained gender-blind. Second, most of those who researched gender in migration, focused mainly on gender and/or women with little incorporation of class, ethnicity, or race as relevant axes of differentiation. Finally, few of these studies included men.

The third stage advanced understanding of gender as a constitutive element of migration. There was an increased recognition of the way gender permeates practices, identities, and institutions in migration-related processes: "Not only is migration shaped by gender relations, but perhaps more important, the migration process experienced by those who pursue family stage migration forges new gender relations. In this sense, migration is both gendered and gendering" (Hondagneu-Sotelo 1992: 411). During this third phase, studies argued for the need to conceptualise gender as one of the constitutive elements of migration processes, including the recognition that these processes themselves may lead to a change in gender relations. The focus on gender then moved to a recognition of the importance of other differences, such as race, ethnicity, and class (Bastia 2014). These studies drew on the wider critiques in feminist theories that argued that race and class together with gender are crucial in the feminist transformation of knowledge: they are "inseparable determinants of inequalities" and "race and class are as significant as gender" (Chow 1996: xix-xx). Studies adopting this approach are in favour of conceptual frameworks that do not take gender as the most important concept for analysing the migrants' experience, but are based on more inclusive frameworks that tend to disrupt the liberal feminist hierarchy of placing gender as the most significant factor (Pessar 1999).

Western feminism has come under increased criticism for its ethnocentrism and the great disregard for issues of race, ethnicity, and class (Benería and Sen 1982). Class opposition is often evident in the way women organise around different issues and specifically in working class women's rejection of middle- and upper-class feminist movements, and vice-versa. Women's interests are clearly often divisive and challenge feminist notions of a global 'sisterhood', such as in the case of the oppressive relations between women domestic workers and their female employers (Benería and Sen 1982). Authors working from an intersectional approach therefore support the notion that issues of gender are intrinsically linked to those of ethnicity, race, and class (Anthias and Yuval-Davis 1983). They argue that these frames of difference should be used together in any theoretical framework, despite the fact that the relative importance of one or the other issue will vary according to the aim of the analysis as well as the context. The same point is reiterated in the work of feminist geographers, who argue for the use of race, class, and ethnicity in conjunction with gender for studying certain issues. Gregson et al. (1997) produced a chronological account of the way feminist geographers have worked with gender. They argued that there were four main stages:

(i) Making women visible, where for example, women's work was highlighted due to male bias prevalent in various social science disciplines; (ii) Gender roles and gender relations, where feminist researchers acknowledged the importance of working with men in order to understand the situation faced by women; (iii) Gender and other social differences, where feminist geographers acknowledged "a hierarchy of social differences with gender at the apex"; (iv) Decentring and destabilising gender, where post-colonial and post-structural feminists questioned gender as the most important analytical category (Gregson et al. 1997: 71).

In a similar vein, West and Fenstermaker (1996) argue that feminist research, suffering from white and middle-class bias, needs to "reconceptualise *difference* as an ongoing interactional accomplishment" (358). They go on to argue that gender cannot effectively be separated from race or class and that research should aim to "build a coherent argument for understanding how they work simultaneously" (West and Fenstermaker 1996: 367).

I started this research by looking at gender issues in a particular migration system. However, it later emerged through fieldwork and data collection that it was impossible to analyse issues related to gender relations without also including race, ethnicity, and class. This shift was well supported by the theoretical work undertaken by various authors who have questioned 'Western feminism' and the related use of the gender variable to the exclusion of issues of race and class (see contributions in Afshar and Maynard 1994; Mohanty 1988). Hooks famously asked: "Since men are not equals ... which men do women want to be equal to? Do women share a common vision of what equality means?" (hooks 1997: 23).

Intersectionality[2]

The differentiated position of women within and between countries in relation to class, race, and ethnicity calls for an intersectional approach to the analysis of gender dynamics in migration processes. While feminist geography has addressed the interconnections of gender with race, ethnicity, and class, as discussed in the previous section, it has begun addressing intersectionality only recently (Valentine 2007). This framework proposes that the various categories of oppression are understood as interconnected and interdependent, rather than as separate essentialist categories, given the limitations of privileging one system of oppression over another and the impossibility of explaining inequalities through a single framework of oppression (Valentine 2007). As Squires argues:

> Theories of intersectionality hold that discrete forms of oppression shape, and are shaped by, one another, and a failure to recognise this results in both simplistic analyses and ill-conceived policy interventions. This approach still retains a notion of structural inequality and operates with groups as the subjects of equality policies rather than individuals, but is

attentive to the cross-cutting nature of structures of oppression and the overlapping nature of groups. (Squires 2008: 55)

Intersectionality is an approach, rooted in feminist theory of power and difference, that is credited to Crenshaw (1991) and critical race theorists on whom she drew, who rejected the notion that class, race, and ethnicity are separate essentialist categories (Davis 1981; Lorde 1985; Crenshaw 1991; Hill Collins 2009). Crenshaw (1991) and others who later took up the term, use intersectionality to draw attention to the interconnections, interdependence, and 'interlocking' of these categories of disadvantage (Brah and Phoenix 2004; Burman 2003; McCall 2005; Valentine 2007). Intersectionality was proposed as an alternative approach aimed at challenging the essentialising notions present in identity politics, which implicitly assumed that "white, middle-class women or black men [are] the exemplary victims of systems of sexism or racism" (Prins 2006: 278). Such essentialising notions inevitably led to the obscuring of intra-group differences: "[i]n the context of violence against women, this elision of difference in identity politics is problematic, fundamentally because the violence that many women experience is often shaped by other dimensions of their identities, such as race and class" (Crenshaw 1993: 1242).

Intersectionality as an approach therefore aims to analyse how different forms of disadvantage intersect and thereby explain the specific experience of certain groups of women on the basis of gender, race, and class simultaneously. Empirically therefore, the intersectional approach aims to highlight the interconnections of the multiple sources of women's oppression and focus on the experiences of those who might have been excluded from feminist analysis (Anthias and Yuval-Davis 1983; Nash 2008).

Conceptually this framework proposes that the various categories of oppression are understood as interconnected and interdependent, rather than as separate essentialist categories, given the limitations of privileging one system of oppression over another and the impossibility of explaining inequalities through a single framework of oppression (Valentine 2007). As Matsuda (1991) argues, it encourages us to ask 'the other question': "When I see something that looks racist, I ask 'Where is the patriarchy in this?' When I see something that looks sexist, I ask 'Where is the heterosexist in this?'" (Matsuda 1991: 1189 – cited in Prins 2006: 279).

Intersectionality aims to destabilise discrete forms of oppression and individual categories of subordination with the aim of exposing their complexity and their interconnections.

Partly as a result of the original work by Crenshaw (1991) but also the ways in which the term was later taken up, race and gender are usually understood as being the 'quintessential intersection' (Nash 2008: 1). This can be explained by the context within which the term emerged: the late 1980s, at the time of the emergent critiques by the black feminist movements in the US and Britain of mainstream feminism as privileging the experiences and interests of white, middle-class women (hooks 1999; Collins 1990). This was also the time of various feminist movements in different parts of the world, closely related to post-independence nation-building in many Asian and African countries and struggles for democratisation and opposition to Structural Adjustment Programmes in the Latin American region (Wieringa 1995). However, Western-based feminist theories have only paid scant attention to the experiences of women outside of Europe and the US. When they have done so, they have often undermined the achievements of non-Western women's movements or their different conceptualisation of women and/or gender. This is also true of those who have adopted and used intersectionality.

The broader context of its emergence within feminist theory was clearly also related to the post-Marxist cultural and identity turn in feminist theories (Fraser 2007). Identity politics emerged as a result of the class-based movement's failing to be inclusive and its promoting the interests of a particular group of workers (Valentine 2007; Young 1998). Postmodern feminism further destabilised the notion of privilege and oppression (hooks 1999; McEwan 2001). These debates challenged the centrality of white, Western, middle-class and heterosexual women and led to pluralist feminist thinking and positions (Rose 1993; Valentine 2007). It highlighted the interconnections and how different systems of oppression feed off of one another.

The politics of difference in feminism coincided with a shift in focus towards the recognition of sexual difference and the deconstruction of masculinity and femininity. Social struggles became subordinated to cultural ones, or what Fraser (2007) describes as a shift from (economic) redistribution to (cultural) recognition. Fraser (2007) argues that this shift was double-edged: it broadened the feminist agenda, but it also weakened the struggle for egalitarian redistribution. It is on this basis that Fraser goes on to propose a broad analysis of gender, which encompasses cultural and socialist-feminist concerns and a broad concept of social justice; "a non-identitarian account of recognition, capable of synergizing with redistribution" (25).

While tracing its roots to the black feminist movement, intersectionality has been taken up more recently by those studying migration, particularly in Northern European contexts. There are clear parallels between the experiences of 'minority' and 'migrant' women. Moreover, intersectionality has been able to address some key concerns in the migration literature.

Following the expansion of intersectionality in the European context there has been a proliferation of studies on migration using intersectionality as their main analytical framework (Lutz et al. 2011). For example, intersectionality has been used to highlight the interplay of categories of difference and identity in Bulgarian migration to Austria (Ludvig 2006). Another study compares the systemic and constructionist approaches to intersectionality through life story narratives of former classmates of Dutch and Moluccan descent (Prins 2006). Ethnic minority women's experience of service provision following domestic violence have also been studied through the intersectional lens (Burman 2003). The relative roles of female, religious, and political identities are explored through the experience of a Dutch politician of

Moroccan descent (Buitelaar 2006), while others use intersectionality to address issues of sexuality in migration studies (Kosnik 2011).

By applying an intersectional approach, these studies begin to redress some key shortcomings in the migration literature. First, by highlighting the importance of disaggregating our analytical categories to reveal important differences, they demonstrate the importance of intra-group differences. For example, one article uses intersectionality to analyse migrant political participation in London (McIlwaine and Bermudez 2011). The authors find that among Colombians in London, working-class women are more likely than working-class men or middle-class women to be involved in politics. On this basis, the authors argue that working-class women are much more likely to challenge gender regimes than middle-class women or working-class men, who are least likely to be involved in politics given their greater desire to return to Colombia.

Second, some authors have taken the concepts advocated in intersectionality and applied them to the study of privilege. For example, a study based on participatory action research with skilled women migrants in Switzerland analyses the barriers they face in accessing skilled employment (Riaño 2011). The author argues that intersectionality is just as useful in understanding privilege, as it is in drawing attention and analysing disadvantages. Riaño's research also reveals the importance of place. She finds that the ways in which the nation state frames different types of migrants has critical consequences for the women she interviewed. In particular, women's married status, combined with their migration status, provided unexpected outcomes, when a highly skilled migrant failed to find a job to suit her skills, while a young single mother with refugee status was provided with free training and childcare which eventually enabled her to go on to hold a professional job. Through her participatory research methodology, Riaño highlights the importance of including both advantage and disadvantage to reveal how gender, class and in her case, nationality intersect – a finding that is also supported by others (Kynsilehto 2011).

Emergent critiques of intersectionality

Despite the uptake and spread of intersectionality in feminist theory and the contributions it made to other areas of inquiry, such as migration studies, it has become apparent that intersectionality theory also has a number of weaknesses. The lack of a clear methodology is clearly one of the main challenges of intersectionality, but this is not the only one. Emergent critiques of intersectionality can be grouped into four main categories: (i) methodology; (ii) scale; (iii) the perceived binary between structure and identity; and (iv) conceptualisations of power.

One of the main critiques of intersectionality is that it does not have a particular methodology (McCall 2005; Nash 2008). McCall (2005) was the first to draw attention to the limited discussion given to methodology. This argument was also supported by Nash (2008), who makes the case that the additive approach to inequality replicates the approaches that are critiqued by those advancing intersectionality, as, for example, when black women who recur to equality legislation need to choose whether to seek protection and redress under gender or under race equality legislation. Nash argues that Crenshaw's approach does not allow for an inclusion of an examination of 'multiple burdens' (2008: 7). She challenges the theoretical reliance of intersectionality on black women's experience on the basis that black women are treated as a homogeneous, unitary social group. For Nash (2008), intersectionality continues to rely on binary identities to explain multiple forms of discrimination. The additive tendency reflects the relative lack of a specific methodology associated with intersectionality.

In terms of methods, on the other hand, the life story approach seems to be the most closely associated with intersectionality. For example, Valentine's study of 'intersectionality as a lived experience', in which she details chronologically and biographically the story of a D/deaf woman in relation to her identity and position vis-à-vis ethnicity, class, and gender, is particularly interesting (Valentine 2007). The analysis shows how Jeannette saw herself differently in different spaces: school, home, and Deaf club. It also shows how "specific spaces (home, work, community) are produced and stabilised by the dominant groups who occupy them, such that they develop hegemonic cultures through which power operates to systematically define ways of being, and to mark out those who are in place or out of place" (2007: 18). Valentine argues that "concern with these broader social structures (and indeed their intersection) has been somewhat forgotten by academic feminism following the theoretical turn within the subdiscipline" (2007: 18) of feminist geography. She aims to readdress attention on the social structures – and their intersections – that produce disadvantage, by analysing how a person's multiple and situated identities place her in a position of disadvantage. Parts of her argument might be specific to debates within feminist geography. However, the approach of focusing on one individual story, while able to illustrate how individuals negotiate (space and) different identities, is limiting in terms of its generalisability and what it can tell us about the ways in which different individuals with different identities and positions in society negotiate power. Many of the studies on migration mentioned above are also based on single life stories. These studies would fall under what McCall (2005) terms 'intracategorical complexity' or the focusing on the complexity of lived experiences in marginalised groups.

Besides methodology, a second major critique of intersectionality is the issue of scale, both in terms of who is deemed to be intersectional as well as in terms of obscuring intra-group differences. While the initial focus on gender and race highlighted the multiple oppression experienced by black women, the focus on this group of women also obscured intra-group differences (Nash 2008; Squires 2008). The question here returns to the subject of "who is intersectional?" (Nash 2008: 9). Nash (2008) asks whether all citizens are intersectional or whether only those who are marginalised have an intersectional identity. Most scholarship focuses on the latter, but some have argued that intersectionality's most significant contribution has been its general theory of identity (Zack 2005). However, if intersectionality is to work as an anti-exclusionary tool, then it needs to address both privilege and oppression and how different axes of differentiation work through each other to produce both: "Progressive scholarship requires a nuanced conception of identity that recognizes the ways in which positions of dominance and subordination work in complex and intersecting ways to constitute subjects' experiences of personhood" (Nash 2008: 10).

If intersectionality is only to focus on disadvantage, oppression, and subordination, as illustrated above, then only those groups who fit these categories are deemed to be the subjects of intersectionality. However, if gender, race, and class (among others) are understood as being relations of power that lead to unequal outcomes, then those who benefit from these social relations and their intersections, could equally be seen as intersectional subjects. This is not to highlight any benefit or disadvantage in a simplistic zero-sum game but rather to recognise that gender, class, and race are not just constitutive of disadvantage, oppression and subordination but also the means through which some people acquire and maintain positions of privilege. In fact, according to some, intersectionality can be applied to any group of people, whether they be advantaged or disadvantaged (Anthias 2002; Brah and Phoenix 2004; Maynard 1994; Yuval-Davis 2006). For Yuval-Davis "[t]his expands the arena of intersectionality to a major analytical tool that challenges hegemonic approaches to the study of stratification as well as reified forms of identity politics" (2006: 201). Privilege, however, does not usually feature in studies of intersectionality (for exceptions, see Kynsilehto 2011; Riaño 2011 mentioned previously).

Besides the scope and focus of analysis, there is also little agreement on the terminology that should be used to refer to the point at which these multiple forms of oppression or privilege intersect. Crenshaw (1991) originally talked of intersections, which is what gave rise to the term 'intersectionality'. However, the road intersection metaphor widely adopted, has not been found to be very useful by others (Yuval-Davis 2006). This is because it seems to suggest that there is an additive element to the different forms of oppression, as in the term 'triple oppression', which does not explain how the different forms of oppression are constitutive of each other (see Anthias and Yuval-Davis 1983). Instead, Yuval-Davis (2007) adopts the language of 'vectors' and argues that "although each of these vectors has a separate ontological basis, in any concrete reality, the intersecting oppressions are mutually constituted by each other. There is no meaning to the notion of 'black', for instance, which is not gendered and classed, no meaning for the notion of 'woman' which is not ethnocized [sic] and classed, etc." (Yuval-Davis 2007: 565).

One problem with the additive approach to equality – where groups are the subject of equality and each strand of inequality is seen as distinctive – is that it limits the development of an integrated equality agenda, such as the recognition of ethnic or religion minorities working against gender equality (Yuval-Davis 2007). One way of dealing with this is by looking at 'complexity within groups' or 'minorities within the minorities' (Yuval-Davis 2007: 56).

The application of intersectionality often resulted in an additive approach to understanding oppression (Anthias and Yuval-Davis 1983; Yuval-Davis 2007). The terminology used to talk about intersectionality discussed above suggests that its understanding of oppression is often thought of in terms of different forms of oppression adding to each other (Anthias and Yuval-Davis 1983; Yuval-Davis 2007). This approach often ends up focusing on the characteristics of the subjects in question, rather than the structural factors that create inequalities (Squires 2008).

In terms of its methodology, this tends to draw the analysis towards the micro level, and to take into account individuals' multiple identities. However, this approach also moves attention away from the structural factors that lead to both oppression and privilege, as discussed above. Moreover, it can lead to the multiplying of different forms of 'oppression' or 'disadvantage'. Is intersectionality only about gender, race, and class? What about sexuality, disability, religion? Many of the studies conducted more recently suggest the latter, as they have added disability and sexuality (Valentine 2007); nationality and ethnicity (Yuval-Davis 2007); migration condition and transnationalism (Chow 2011) to the original categories of differentiation (race, class, and gender). This highlights the vagueness of the term, which Nash (2008) argues is also one of the strengths of intersectionality. However, this tendency to broaden the focus of intersectionality needs to be addressed critically. Why is there a need to broaden out its focus? In addition, does this broadening also imply a danger of weakening its analytical potential?

Fraser (2007) proposes a broader analysis of gender that considers the original socialist feminist concerns about distribution with the newer concerns of identity and culture. She does this by first widening the concept of justice to encompass both distribution and recognition "and a non-identitarian account of recognition, capable of synergizing with redistribution" (2007: 23). While widening and enriching the feminist project, Fraser (2007) argues that the focus on identity and culture can work to displace the original focus on redistribution in the current neo-liberal context. The 1970s socialist feminist project conceptualised within a political economy framework gave way to psychoanalytical identity based theories which rejected the term 'gender relations' as too sociological and replaced it with 'sexual difference', a term which is more closely linked with subjectivity and the symbolic order (Fraser 2007). By the 1990s, most feminists worked within the cultural turn on issues of gender as identity and as culturally constructed (Braidotti 1994; Butler 1990). On the basis of the variety of studies that have taken up intersectionality, it seems clear that the theory can be used in either approach: one based on identity or one based on structural inequalities. What cannot be expected of intersectionality is that it resolves this wider tension present in feminist theory.

Finally, many have argued that intersectionality does not really offer a new theory of power. While scale and the (false) dualism of structure vs. identity represent one of the main tensions in studies about intersectionality, its limited theorisation of power seems to be its main weakness. Nash (2008) questions whether intersectionality contributes any new tools to black feminism to develop a more complex theory of identity. Given the lack of consensus that identity should be the focus of feminist research, the question needs broadening out to include a more general theory of how power is organised within society, and needs to include social structures. Intersectionality does not provide any new tools as such. Its usefulness lies in bringing existing concepts, such as gender, class, and race, into the same analytical framework to try to understand how they relate to one another. However, as argued above, this approach needs to be rooted in a historically grounded and context-specific analysis of social relations of difference in order to avoid depoliticising and simplifying complex realities.

In this book, I have therefore taken up the invitation proposed by intersectionality theorists to help frame the analysis of social change in relation to gender, class, ethnicity, and race but within a socio-historical framework of what these terms mean in the context within which migration takes place. I have not applied a firm framework as such because in my understanding of intersectionality, this does not exist. I therefore offer this analysis as a response to a specific invitation to explore the intersections of gender, race, ethnicity, and class, and destabilise the primary importance of gender as *the* most important axis of disadvantage – but without losing sight of intersectionality's feminist origins.

Transnationalism

The feminist literature on gender and migration shows a definite tendency towards exploring changes at the individual level, that is, the ways in which migrants themselves experience these changes. While this in itself is valuable, there is also a need for a more thorough inclusion of meso levels of analysis. Migration theory is best suited to a meso-level analysis, but one that also includes the necessary recognition of changes that take place at both the macro and the micro levels (Castles 2010). Transnationalism is one such approach that emerged over 30 years ago, but nevertheless proposes the inclusion of the analysis of global transformations in capital relations as well as placing recognition of the centre-stage importance played in migration by social networks. It is particularly useful for the purposes of this book because it allows for the linking of changes at the macro-level, such as globalisation and economic crises, with the micro level, of how individuals react and make decisions in relation to these changes, passing by the meso levels of functioning of households and labour markets.

Transnationalism highlights the interconnected nature of these global processes and focuses our analytical gaze on the social relationships sustained across places of origin and destinations. The framework is based on four premises and argues that migration is linked to global capitalism and as such it must be analysed within the context of global relations between capital and labour (Basch et al. 1994). It introduces the concept of the 'transnational social field', i.e. the social fields migrants create across national boundaries, and argues that transnationalism cannot be analysed through bounded social science categories that conflate physical location, culture, and identity. The analytical framework also introduces the notion that transnational migrants contribute to the construction of two or more nation-states. Taking these four elements into account, Basch et al. (1994) argue that today's migrations are substantially different because of the current moment of capitalism that places migrations in a vulnerable situation, one in which migrants have little or no job security while facing continued racism. Job insecurity and widespread racism in turn place migrants in a situation where they need to maintain and cultivate their fall-back position, offered by family members and acquaintances who remain in the countries of origin. In this process the inter-dependencies that have developed between nation-states as a result of the movement of labour in one direction and the flow of financial remittances in the opposite direction (Phillips 2009) is also experienced at the micro and meso level between households and communities that stretch across national borders.

Although transnationalism has evolved considerably since first developed by Basch et al. (1994) (see also Portes et al. 1999), it presents a useful and, at the time, novel way of studying migration precisely because it was developed out of an appreciation of the consequences the global expansion of capital was having on migrants' different forms of belonging, including issues of labour market integration, political participation in the country of origin, and destination and the issue of return. In addition, while initially somewhat overlooked, people working on transnationalism have also embraced the challenge of integrating gender relations within this framework (Mahler and Pessar 2006). This is critical, not only because of what has become known as the 'feminisation of migration' but more importantly, because, as already discussed, gender relations permeate all social relations, and therefore, are also fundamental for the ways in which the process of migration develops, how it is experienced, and the consequences that migration can have.

Transnationalism is also relevant for how we think about issues of justice. While historically justice was assumed to be dealt with at the national level (Fraser 2005), the current historical junction as well as transnational living practices require us to rethink the level at which issues of justice are dealt with. Within this expanded framework of global justice, migration can be understood as a strategy employed by individuals, households, and sometimes whole communities to get access to justice. People may leave their places of origin because of political persecution and their inability to participate equally in taking political decisions. Fraser terms this process 'misrepresentation' and more specifically 'misframing', when the boundaries of a political community are set in such a way as to wrongly deny some people the possibility of equal participation (Fraser 2005). Others leave for higher income countries in search for better paid jobs and new opportunities for upward social mobility. Their movement to the global as well as regional centres of economic power, and the remittances they send to their countries of origin, represent an economic element of justice, or what Fraser terms 'redistribution' (Fraser 2007). Finally, many seek new lives abroad as an avenue for achieving modernity, inclusion in a global citizenship, shedding their rurality and status as second class citizens, aspiring to become modern, urban cosmopolitans (Kothari 2008), or what Fraser terms 'recognition', the cultural dimension of justice (Fraser 2007).

These different dimensions of migration are not necessarily as clear-cut as outlined and they often overlap, as do the structural and cultural dimension of different forms of oppression. Yet it is useful to distinguish them at this stage because of their multiple interconnections. For example, women may cite redistribution and their inability to make ends meet with local wages as the main reason for seeking work abroad. However, they may want to move abroad in order to leave an abusive partner or an unhappy marriage in contexts where it is socially unacceptable for a woman to seek separation from her husband, as found in Sri Lanka (Gamburd 2000), the Philippines (Gibson et al. 2001) and others researching gender and migration in Asia (Piper and Roces 2003). Migration in this case is therefore an act of resistance towards the fundamentally unjust ways in which gender relations are structured in the society in which they live. This is further illustrated by different attitudes towards return as expressed by men and women. Evidence, particularly from the US, suggests that women generally do not wish to return to their countries of origin because they achieved a higher status in the country of destination. The opposite is generally true for men, who wish to return to their countries of origin where they are able to benefit from a relative higher status (Donato et al. 2006; Hondagneu-Sotelo 1994).

Nations Unbound was written on the basis of fieldwork carried out throughout the 1980s, at the time of the debt crisis, Structural Adjustment Programmes, and retreating gains from the welfare states founded in the post-WWII period. In defining the focus of the book, the authors came to the conclusion that they will "develop a framework that could encompass both transnational practices and the contention about the identities of the transmigrants taking place within the United States and post-colonial nation-states" (Basch et al. 1994: 22). Despite later critiques accusing *Nations Unbound* of methodological nationalism, the authors are clear in their critical approach towards the nation state, as always differentiated on the basis of class and race (and gender, although less is made of this point in the book). They cite Hall (1977) to support their argument that "unity may be secured 'in and through (not despite) differences'" (Basch et al. 1994: 35). In addition, their focus on identities is complemented and in fact intrinsically linked to their acknowledgement of the ways in which transnational ways of living were at the time a necessity given the increasing difficulties (economic, social,

political) that migrants faced both in countries of origin and those of destination, concluding that: "Economic and political vulnerability, magnified by
the factor of race, augment the likelihood that migrants will construct a
transnational existence" (Basch et al. 1994: 27).

My own research was initially framed by the largely US-based sociology of
gender and migration, but I became interested in transnational practices given
my empirical observations of the ways in which the community where I
decided to do my fieldwork effectively crossed national borders and involved
multiple, mostly urban, spaces that belong to different nation states. The
penny dropped when a story I had heard in Buenos Aires about the death of a
young Bolivian woman in a motorbike accident mentioned in the introduction was recounted with exactly the same details a year later in Cochabamba,
Bolivia, in the neighbourhood where her father lived. When listening to this
story in both locations, Buenos Aires and Cochabamba, I was struck not just
by the similarities in the details, but also by the shared emotional references
used to describe what had happened and the ways in which the 'community'
mobilized to help her father travel from Cochabamba to Buenos Aires to
recover her body.

My research interest then and to some extent still now, was on the ways in
which gender relations influence and are influenced by migration. I realised
early on that the 'nation state' or Bolivia, as a unit of analysis, would not take
me very far given the multi-ethnic constitution of the Bolivian 'nation' in
which the only 'shared' language is the one that was imposed during colonial
rule[3]. Despite the many problems inherent in the concept of 'community', I
decided to frame my research within a particular 'community', which I
sometimes also describe as a neighbourhood.

The emancipatory potential of migration[4]

In an increasingly globalised and interconnected world, it is imperative that
we pay attention to whether existing inequalities are exacerbated or challenged by the factors associated with globalisation and the so-called network
society, such as migration. For a feminist geography of migration, therefore,
the challenge is to place women at the heart of our inquiry to better understand the extent to which they may benefit from their migrations. Feminist
theories are well-placed to help frame such an inquiry (Hondagneu-Sotelo
2000; Bonifacio 2012).

Migration studies have recently experienced a resurgence of interest in the
question of migration and social change (Van Hear 2010; Castles 2010; Portes
2010). Yet despite repeated calls for the need to integrate an analysis of
gender, and other social inequalities, into migration studies, mainstream
scholarship continues to address migration as a gender-neutral process. The
special issue on social transformation referenced above, for example, pays
little attention to how gender might be an integral form of the changes
brought about by migration, with the exception of the one paper that deals

explicitly with gender and migration (Lutz 2010). The challenge from a feminist perspective, is to integrate a gender-sensitive analysis into all studies of migration.

Feminism is inherently interested in promoting gender equity and achieving an equitable society where women and men enjoy not only the same opportunities but also equal status and worth. In migration, feminism helped shape an understanding of migration in which gender and migration are understood as constituting each other, i.e. representing a dialectical relationship in which gender relations influence migration and are in turn influenced by migration (Hondagneu-Sotelo 1994; Kofman 1999; Willis and Yeoh 2000; Pessar 2005; Curran et al. 2006). Moreover, following the critiques of the black women's movement, post-colonial critics, and those who have historically been misrepresented by feminist knowledge and scholarship, it is clearly important to acknowledge the multiple forms of feminist organising as well as multiple axes of disadvantage. This leads us to destabilise gender as necessarily the most important or relevant basis of unequal power relations (McEwan 2001; Radcliffe 2006; Mohanty 1988). Some of these critiques have been taken up through the use of intersectionality, although this also carries its own set of problems, as already discussed. It is clear that migration has the potential for promoting progressive social change. As migrants move from one place to another, whether within the same country or across national borders, they experience new cultures and ways of being, that can lead them to question their own culture and society, thereby leading them to challenge the status quo (Bastia 2011). However, most evidence suggests that migration reinforces existing social structures. De Haas (2012) for example, argues:

> Despite the often considerable benefits of migration and remittances for individuals and communities involved, migrants alone can generally not remove more structural development constraints and migration may actually contribute to development stagnation and reinforce the political status quo. (8)

This can be verified empirically, but is also a logical conclusion if we consider what we already know about migration: that the poorest generally lack the human, financial, and social capital required to migrate (Skeldon 1997; Kothari 2002, 2003). Those who are able to migrate are therefore generally better off, if not the elite, who then reap even further benefits through migration and reinforce existing inequalities. Moreover, migration itself may also be an escape valve for societies in turmoil: by emigrating, migrants may be contributing to decreasing the overall levels of unemployment as well as weakening social and political demands on the state (Glick Schiller 2011; De Haas 2012). However, evidence to date is inconclusive and largely context-specific (Black et al. 2005; Bastia 2013b).

For gender relations, the causal relationship is even more complicated. While migration per se has the potential to disrupt existing gender-based ideology, it is unlikely to do so in the long term because women often trade any potential for disrupting gender relations for upward class mobility, a strategy they share with their partners (Bastia 2011a). In the context where women 'stay behind', i.e. where they do not migrate, they may engage in new activities as a result of their husbands' absence, but these may be seen as a burden rather than a source of empowerment (De Haas and van Rooij 2010).

In some of the literature on gender and migration, there is an underlying assumption that migration can lead to more equal gender relations, either through increased women's labour market participation, higher income, or through the impact of more 'modern' and liberal models of gender relations. Many of the studies arriving at such conclusions are framed within a liberal feminist framework, one that sees women's entry into paid work as a precondition for achieving gender equality and women's liberation (Arruzza et al. 2018). Such framings also often assume that gender relations in women's countries of origins are necessarily more restrictive and oppressive, as compared to those of the richer countries that women migrate to (Bastia 2011a).

Migration clearly represents an opportunity for forging new social relations as migrants, the social actors, leave their communities of origin, and establish new social relations through the migration process. Such abrupt changes enable social actors to start questioning social relations and modes of behaviour that have until then been part of the "universe of the undiscussed (undisputed)" and what Bourdieu terms *doxa* (Bourdieu 1977). Confrontation with other ways of organising social life, including power relations between women and men, could challenge established ways of doing things. This process would bring inequalities into the realm of *opinion* "the locus of the confrontation of competing discourses – whose political truth may be overtly declared or may remain hidden, even from the eyes of those engaged in it, under the guise of religious or philosophical oppositions" (Bourdieu 1977: 168).

In practice, evidence on the 'emancipatory' role of migration is sketchy and contradictory. Early studies of Turkish women migrants found that migration to Germany led to the emancipation of those women who migrated and entered the labour market in Germany as well as those who remained behind while their husbands worked abroad. Although to different degrees, both groups of women had greater access to income as well as greater decision-making power within their families. In addition, migration led to a change in the traditional family structure, with a greater acceptance of nuclear families and a loosened control of young, unmarried women (Abadan-Unat 1977). A collection of case studies from Europe, Africa, Asia, and Latin America, illustrates the inconsistent nature of changes in gender relations that take place as a result of women and men engaging in migration (Buijs 1993). For example, Palestinian women refugees in West Berlin found their freedom restricted by an imposition of domesticity and increased control by male members of their communities (Abdulrahim 1993). On the other hand, Chilean women refugees in the US have experienced increased independence and self-esteem by their newly-acquired skills and labour market participation, which made them reluctant to return 'home'. This was in contrast to their male counterparts, whose self-esteem was lowered due to their loss of labour union participation and therefore part of their identity, which was central to their exile (Eastmond 1993).

Internal migration can similarly lead to significant changes in gender relations. A study of internal migration in Albania found that the patriarchal organisation of rural households starts to break down through migration (Çaro, Bailey and Van Wissen 2012). Internal migration in Albania is often initiated by women and then developed as a family project, unlike international migration, which is initiated by men. Migrant women associated internal migration with modernisation, such as access to modern clothes, make-up and hair styling but also a move from extended to nuclear families. Women who had migrated internally were more likely to have paid work outside the household and engage in what the authors call 'female networking', or activities with other women, for which they just inform their husbands, instead of asking for permission (Çaro, Bailey and Van Wissen 2012).

Others have suggested that migration is more likely to be empowering if some of the following elements are present in a specific migration stream: the migration is rural to urban; it is not undocumented; women work outside the home at destination; women move autonomously and not as part of a family group; women enter formal sector occupations; and migration is longer term, not temporary (Hugo 2002: 27). In addition, having a different set of relationships with other women at destination can also increase the potential empowering effect of migration (Hugo 2002). This is clearly not a set of criteria as such, but following on from Hugo's suggestion, one would expect that migration from the Cochabamba neighbourhood to Argentina and Spain not to be significantly empowering for women. Although part of this migration has rural origins, the transnational migration is urban to urban, it is mostly undocumented and temporary; and while women migrate autonomously, or at least some of them do, women generally do not enter formal occupations. However, to better understand the nature of this migration and assess to what extent it has been empowering, my research has focused on the ways in which women experienced autonomy: in relation to their mobility and through their use of social networks, in the types of jobs they accessed, in the ways in which care and social reproduction have been reorganised as a result of migration, and in their intimate lives.

Autonomy[5]

Some analysis has used autonomy to gauge changes in gender relations. During the second wave of the feminist movement, autonomy was seen as a precondition for achieving gender equality (de Beauvoir 1993). However, more recently, feminist theorists, including feminist geographers have become more ambivalent towards the concept of autonomy (Bondi 2004; Mackenzie and Stoljar 2000). Part of the ambivalence expressed towards the concept of autonomy stems from the conflation of autonomy "as a means to an end or an end in itself" (Bondi 2004:10) and the fact that it can be mobilised for the pursuit of rather different ends, namely independence and *separatedness* (Kabeer 1999; Morgan 1970).

Autonomy can be seen as breaking away from normative expectations of gendered behaviour and a step towards achieving gender equality. In this sense, autonomy indicates an open rejection or at least a challenge to patriarchal control. However, autonomy is usually seen as less desirable than achieving systemic transformation and gender equity (Kabeer 1999), given that the concept of autonomy draws on androcentric notions of atomised, individualised behaviour, and devalues relatedness, interdependence, and social aspects of people's lives (see e.g. Jackson 2003; Mackenzie and Stoljar 2000). These authors therefore reject autonomy because it is linked to a particular idea of autonomy, that of the "caricature of individual autonomy as exemplified by the self-sufficient, rugged male individualist, rational maximizing chooser of libertarian theory" (Mackenzie and Stoljar 2000: 5). If we recognise that choices and the person making them are not only embedded in wider social relations and power structures, but also, "*constituted* by them" (Hirschmann 2003: 204, emphasis in the original), we realise that these choices and the people making them could not exist outside of social relations.

Like Hirschmann (2003), I understand autonomy as related to capabilities and women's ability to make choices about their own lives: "[a]utonomy is fundamentally about capabilities, specifically about the ability to assess one's options, reflect critically about them, and make choices that allow one to exert some control over one's life" (36). In this sense autonomy is closely associated with freedom and liberty, but includes the possibility of understanding autonomy as embedded within everyday social relations. Hirschmann (2003) then goes on to focus on freedom: not because she rejects the concept of autonomy but because "freedom prompts somewhat different questions" (39). However, I would argue that once placed within the context of embedded social relations and power inequalities, the term can regain its usefulness (Mackenzie and Stoljar 2000), particularly because autonomy has the potential to shake and disrupt patriarchy (Daya 2009; McIlwaine 2010; Song 2010).

Autonomy is useful in analysing the ways in which migration can potentially lead to broader social change in gender relations. I therefore use it in conjunction with attention to the meso-level analysis mentioned earlier, which is best suited for migration studies. I begin with a focus on social networks and the extent to which different types of social networks are able to support women's autonomous migrations (Chapter Four), and then move on to the type of work that they do (Chapter Five), the extent to which women are able to renegotiate caring responsibilities (Chapter Six), and their intimate lives (Chapter Seven). Taken together, these four elements: social networks, work, care, and intimacy, provide a broad framework from which to evaluate social transformation in migration or as a result of migration in the migrants' economic, social, and personal lives.

While the focus in these questions is mostly on women, the aim is to understand how gender relations change through migration. I also pay attention to men and their strategies and experiences of migration. However, given the feminist focus in this research, I am interested in understanding how women's position changes within gender relations and whether migration opens up some possibilities for greater gender equality.

Notes

1 Parts of this chapter were first published in Tanja Bastia, 2011, Migration as protest: negotiating gender, class, and ethnicity in urban Bolivia, *Environment and Planning A*, 43(7): 1514–1529.
2 This section and the next were first published in Tanja Bastia, 2014, Intersectionality, migration and development, *Progress in Development Studies*, 14(3): 237–248.
3 However, even indigenous languages have their own history of colonialism, with Quechua being imposed by the Incas on other ethnic groups in their endeavour for expansion and (pre-Spanish) conquest.
4 Parts of this section first appeared in Tanja Bastia, 2013, The migration-development nexus: current challenges and future research agendas, *Geography Compass,* 7 (7): 464–477.
5 This section was first published in Tanja Bastia, 2013, I am going, with or without you: autonomy in Bolivian transnational migrations, *Gender, Place and Culture* 20 (2): 160–177.

3 Origins

Migration patterns vary across time and space. This chapter analyses the characteristics of the community of migrant ex-miners under consideration in this book through a community survey and life stories. The analysis sheds light on the particularities of migration-related characteristics of members of the community of ex-miners who have relocated to the outskirts of Cochabamba as compared to those of Bolivians migrating to Argentina as a whole and migrants in other countries.

Identity represents the basis on which particular relations are forged. These in turn mediate the development of particular processes in this community of reference. This chapter therefore starts from the migrants' immediate origins, the mining town, in order to give 'identity' a more thorough grounding. This is not to say that the mining town is taken as the absolute place of origin for the migrants interviewed. Migration is often complex and rarely involves a single journey from A to B (Skeldon 1997). For many of the migrant ex-miners, the mining town was only one of the many stops they had throughout their lives. Others were born in the mining town but their parents migrated there from other mining towns or from peasant communities. As such, they often continue to maintain relationships to these various places.

Given the dynamism of people's mobility in a country like Bolivia, it is hard to identify a fixed point of origin for any community. It is better to understand the 'community' as being constituted by the social networks its members reproduce in their daily lives. In this sense, 'community' is not geographically bound. It can be constructed and reproduced across different geographical regions and political entities, still maintaining a strong link with a constantly-changing and adapting specific community of 'origin'. However, the time spent in the mining town was often decisive for migrants' identities. It is because of this that this chapter starts by briefly discussing what many consider to be their 'origins'. The migrants' identities as miners, ex-miners, miners' wives, and children of mining communities and former mining communities are also illustrated by the life stories that will be presented in more detail in this chapter.

Tracing elusive origins

Sergio, an ex-miner in his fifties, interviewed in his house in Cochabamba in 2002 began to describe his life in the following way:

> My life in my youth was in the mining centre. I worked in the mines for twelve years, until I was 40 years old I worked in the mine, then I came here. We had bought some land in a group and we came. I have been here since 1989. I was working for a little while in the council of Cochabamba, in the city, the parks and gardens section and then, I had a little problem and I had to go away to earn some money in another country. And I went to Argentina. Then, in Argentina, it went well. (Cochabamba, 9th June 2002)

'Community' is a much debated concept and needs to be treated with reservation. Besides the body of knowledge showing that communities as such are largely 'imagined' (Anderson 1983), the term has also been widely criticised as being used in such a way as to mask internal inequalities while highlighting the 'public face' of a community (Guijt and Shah 1998). However, the concept of 'community' is still useful especially in view of the fact that the migrants interviewed construct their identity as being part of this community whose origin takes them back to the mining town. When this concept is employed in conjunction with the analysis of migratory movements, it can be said that this community can be described for our purposes as a transnational community (Basch et al. 1994).

At this point I need to distinguish between this community and the whole group of reference. While I restricted research in Bolivia to a specific community who traces its origins to the mining town, I diversified data-collection in Buenos Aires to include migrants who were born or brought up in other places but who have in some ways become connected with this core group of ex-miners, through either work, friendship, or marriage/cohabitation. This diversification of the group of reference is useful for the purposes of understanding migration, social networks, and upward social mobility, but the opposite is true in relation to the analysis of community feelings and identity. Therefore, I will make a distinction, where appropriate, between those who identify themselves as being part of this community and those who do not.

For the purposes of this book it is useful to conceptualise the 'community' as being constituted by the social networks its members construct and maintain. As Wellman put it "the trick is to treat community as a social network rather than as a place" (Wellman 1999: xv quoted in Vertovec 2001). This, according to Rogers and Vertovec (1995) allows for social interaction to be conceptualised without having to rely on actual spatial proximity. It follows that the community itself is actually constituted by its own social networks. Without social networks, there is no community. Therefore, when we talk about social networks we are actually talking about the community itself. This conceptualisation still leaves space for reference to be made to a particular geographical space. Given the high degree of spatial mobility shown by the members of this community, the geographical dimension of these social networks itself merits some attention (Voigt-Graf 2004). I discuss these in Chapter Four.

The social networks that migrants use to facilitate their migration are organised on and above the geographical locations where members of this community live and work. These geographical networks include the community of origin and the places ex-miners and their families have migrated to, including, but not limited to Oruro, Cochabamba, the region of Chapare, Buenos Aires, and various cities in Spain. What allows these locations to be interpreted as being interconnected are the social networks that are developed and maintained by members of this community. Social networks are built on the existing relationships within the nuclear and extended families. Besides immediate and distant relations, social networks are also formed based on *compadrazgo* or fictive kin relationships, friendships and ultimately, on community membership.

Social networks have been given a lot of consideration in migration studies since the pioneering research of Barnes (1954) in his study of a Norwegian Island parish. They clearly play a central part in most types of migration, channelling information between the place of origin and destination, providing economic, emotional, and practical support to many migrants. In this community, social networks also form the basis on which migration takes place.

The mining town[1]

The immediate origin of this transnational community is a mining centre, located in the mountainous north-western part of the department of Cochabamba, Bolivia at around 3,800m above sea level. In its heyday, it used to be an important economic centre, not only providing employment for thousands of miners but provisioning the neighbouring rural communities with education and health services. It used to attract people from the neighbouring communities as well as from further away, as Sandra explained:

> Well, I am from the mining centre [...]. My parents are migrants from Potosí and my mother was born in Cochabamba. They went to the mining town from the provinces [rural areas]. They came to the mining town and I was born there, in between the mountains. (Buenos Aires, 23rd February 2003)

A small number of families worked in the service sector, but most families relied heavily on the mining industry for their livelihood. The dominant gender ideology assigned reproductive roles to women, who were brought up to be homeworkers while men, the 'miners', were assigned the breadwinning role. However, families were numerous and mining a dangerous occupation, which led to many miners' premature death due to accidents or illness. This led to some women working in the mining industry as well.

Supra-natural beliefs forbid women from entering into underground mining tunnels. Despite this, some women did work underground, albeit within a strict sexual division of labour. However, these women were few. Most women worked in other occupations, such as that of *palliri,* an occupation usually undertaken by women, involving sorting out rocks containing ore on the waste piles or night watching (Nash 1976; Bocangel Jerez 2001). Nash (1976) conducted research in the Siglo XX tin mines and similarly describes how women usually worked above-ground, except for periods when male labour was scarce, such as during wars.

Living conditions in the mining centre were hard. Miners rarely had access to running water, sanitation, or modern cooking appliances. Working conditions were similarly difficult, especially for women. A national study on artisanal and small-scale mining in Bolivia argues that widowed *palliris* earned no more than 25 per cent of the minimum wage – then set at Bs.400 or US$50 per month – while night watchers earned between 55 and 66 per cent of the minimum national wage. Married *palliris* generally received no remuneration at all, adding their ore to those mined by their husbands' (Bocangel Jerez 2001).

Facing such difficult working conditions and few livelihood alternatives, parents started giving priority to education for both girls and boys born in the 1970s onwards. As in other Latin American countries, education is often seen as the only way out of poverty and for 'becoming somebody' (Crivello 2011). In the past, it was not uncommon for children to start working at an early age and many elder members of the community have very limited schooling. Women over 49 years old have on average only three years of schooling, while men have four years. This started changing as miners lost confidence in mining as a secure occupation and moved to the urban areas.

The literature on transnationalism highlights the importance of a 'place of origin' and argues that as migrants move across the globe fleeing persecution or in search of better opportunities, they seek to reproduce the nation of origin (Basch et al. 1994). A similar process takes place at the level of the community, even when the movement is within the same country.

'Place' is critical for migrants' construction of their own identity, a sense of unity, cohesion, shared history, and solidarity. Here I highlight the use of the word 'a sense'. This is a feeling, an ideal. Reality, as feminist studies have widely shown, is much more fragmented, and intersected with unequal power relations based on gender, race, ethnicity, and class, creating multiple exclusions and different forms of inequalities. Yet migrants create this ideal place of origin because of a need to 'anchor' the physical, social, and cultural displacement they experience in their everyday lives, as internal migrants in their own countries or as international migrants.

The miners I talked to usually identified themselves in relation to the mining town. This identity was so strong that it left little space for negotiating difference and was used to unite this group of people in their difficult process of internal migration, which followed the intense economic crisis of the mid 1980s. The cooperative began subdividing a large farm they had bought on the outskirts of Cochabamba into individual 150 square metre plots and allocating them to those members who had paid a certain amount of contributions. These plots were part of a farm and as such were designated as 'green land', not to be built upon, by the local authority. The neighbourhood struggled for twenty years to convert their status from 'green land' to urban. In the process, they also acquired basic services, such as water and electricity, by private means for the former and from the local authority for the latter. With these land titles the miners began leaving the mining town during the late 1980s and early 1990s to set up a new home on the outskirts of Cochabamba, amidst rising unemployment and increasing economic informality. They remained united physically because of the plot of land that the mining cooperative bought, but their identity also provided them with a sense of unity and a shared history.

However, the mining town was never given. It was constructed through unequal power relations, movements, and flows of people and capital. There are three main axes of differentiation that are relevant to our discussion: ethnicity, class, and gender relations (McDowell 2008). Already during my first fieldwork in 2002/3, some interviewees identified themselves in relation to the legitimacy given to them in light of their birth in the mining town. The 'original' inhabitants of this mining centre differentiate between themselves and those who moved into the community more recently, as is evident in the account of Claudia, who was in her fifties at the time of the interview:

> And then there was the cooperative in the mining town [...] at that moment Sandra's father arrived too. It was when I was small, because they are not from the place. I am from the place, from the mining town. I was born in the mining town. [...] I am a legitimate [person from the mining town] from my birth I am [a person from the mining town]. That is why I know those people who came afterwards. (Buenos Aires, 11th March 2003)

The mining town was made of people who were born in the locality, miners from other mining towns, such as Llallagua and Siglo XX, and peasants who took up temporary mining jobs but ended up staying and contributed to the making of the place. They became part of the mining town and their identity as miners, or miners' children, held throughout their migration history, as a female returnee with migration experience to Argentina and Spain explained:

> I consider myself from the mining town in every sense of the word. I am as any other person who belongs to a place, who adopted its customs, its roots, and its traditions. [...] I was born there, I grew up there. Through that process of growing up I acquired many things, many values or

customs [...] I learnt the good and also the bad. I think we are the product of the environment we were born in and grew up in and it's because
of this for example that I don't want to leave the neighbourhood [in
Cochabamba]. I don't want to leave because I already know my people.
Despite the fact that there might be protests and there might be a
moment in which we fight among ourselves. But that is just a moment in
time. It passes. (Cochabamba, 16th May 2010)

For her the conflict was temporary while her place of birth continues to
represent a fundamental part of her identity. Miners' identity was often constructed in opposition to peasants, whom they saw as being less civilized (Gill
1997). Most of my interviewees spoke one or sometimes two of the main
indigenous languages spoken in Bolivia (Quechua and Aymara). In the
neighbourhood, only 24 per cent speak only Spanish; 64 per cent are bilin-
gual Quechua and Spanish speakers while about 10 per cent speak also
Aymara (Survey 2008). However, the great majority were fluent in Spanish
and they used language as the basis for differentiating themselves from
peasants: "There is a difference between us and the people who cannot
speak Spanish. They are pure Quechuas or Aymaras" (Cochabamba, 16th
May 2010). While recognising their indigenous roots, they do not identify as
indigenous. As others have shown, this is a complex issue in Bolivia, where
the term *indígena* refers to indigenous people in the lowlands. Indigenous
people from the highlights were described in Spanish until the 1952 national
revolution by the term *indio,* which is pejorative and often used as an insult.
Since then, they have been labelled as *campesinos,* literally peasants, which
refers more to their class rather than their ethnic identity (Canessa 2007;
Regalsky 2003).

During my recent fieldwork a man in his forties, who used to work as an
engineer in the mining town compared Bolivian migrants in Spain to peasants
in the mining town. When peasants came to the mining town, miners would
serve them a plate full to the brim with food. He was offended that as a
migrant in Spain, Spanish people would treat him in a very similar way by
serving him huge plates of food (fieldwork notes, 19th April 2010).

A further division was created on the basis of employment, more easily
identified as a class difference, between those who worked for the state-owned
Comibol and those who worked for the cooperative. While both enterprises
dedicated themselves to mining, the cooperative was effectively created when
Comibol ceased to be profitable and leased its concessions to the cooperative.
However, the government retained its presence there and continued doing
explorations as well as setting up an electricity project, so there were a few,
mainly professional workers who were employed by Comibol. Besides receiving a regular income, they were also entitled to large shares in the public
distribution system that was operating at the time, *pulperías.*

Claudia explained the way only those fit to work were accepted into
Comibol and those who were rejected had no choice but to work on their

own. They formed a cooperative, but its members had a much lower standard of living than those who were working for the state agency.

> So those who were in good health entered Comibol. Those who were ill couldn't get in. For that reason, what are people to live on? For that reason they established the cooperative. [...] They had everything. You don't know how the Comibolistas enjoyed themselves. They really enjoyed themselves. They used to have 30kg of meat every 15 days. Then, 100kg of sugar, rice, hmmm. (Buenos Aires, 11th March 2003)

They shared the education and health services with the cooperative workers, so that all children attended the same school. However, this different source of employment created a significant socio-economic difference that was reproduced to some extent through the migration process. For example, some ex-Comibol employees moved directly to the centre of Cochabamba – rather than the plot owned by the cooperative, which is on the outskirts – to pursue higher education. In Buenos Aires, they often moved out of the shanty towns, where most people from the mining town lived.

These differences produced material inequalities and were highly gendered. While the peasant economy was built on women's undervalued yet active participation in the household agricultural economy, miners' identity was constructed on the breadwinner model (Nash 1993). This model was most easily reproduced by the professional class or those working for Comibol than the *cooperativistas* whose income was insecure. However, they both shared the same values of gender relations.

The Catholic Church had a strong presence in the communities, organising craft activities for women, usually in exchange for foodstuffs. Women did not see themselves as being politically involved: "how can I explain, those who knew about politics, they dealt with it and they managed us like little animals. We were followers. They would give us rice, sugar; well … we let them buy us off" (Cochabamba, 25th May 2008). In other mining towns, when women organised politically, at times staging hunger strikes that brought down dictatorships, they still represented themselves as being 'miners' wives' (Barrios de Chungara et al. 1979), despite the fact that the very act of organising would change their lives, as was found elsewhere (Laurie 1999).

Family planning was hard to find and, combined with men's overwhelming power to dictate the nature of sexual relations within married or cohabiting couples, meant that women generally had large numbers of children (see Boesten 2010 for a related discussion but on Peru). An elderly woman explained that she had nine children because her "husband was very jealous" (Cochabamba, 30th April 2008). Young girls also had low levels of control over their own bodies and it was common for their first sexual experience to be non-consensual. Some young girls were abducted and then forced to marry their abductor (see example on page 145). Others were pressured to marry the person they had their first sexual relation with, on the grounds that nobody would want to establish a serious relationship with a girl who was known to have lost her virginity (see Chapter Seven on intimacy).

The ideology of *machismo* is clearly linked to a particular sexual division of labour, the breadwinner model where the man works and the woman is in charge of reproductive tasks, as a man who used to be a miner explained: "in the mine we are *machista*. Why? Because the man works and the woman has to look after the children" (Cochabamba, 14th May 2008). Men used violence as an explicit strategy to control women's behaviour:

> First of all, we were *machista,* the man, well, we used to call ourselves *manmón,* the one who is in charge [*manda*], and the one who mounts [*monta*]. *Manmón* we used to say. We used to say 'You have to beat your woman at least once a week, just in case, with or without reason'. Why? So that she behaves. (Cochabamba, 14th May 2008)

Domestic violence was common and widespread. Even before large-scale migration from the mining town to Argentina began, international migration was related to domestic violence. During the last fieldwork, Claudia, mentioned earlier, who was also one of the first women from the mining town to migrate to Argentina, told me how one of her neighbours went from the mining town to Buenos Aires in 1961/62 because her husband was *malo* or very bad. He used to beat her a lot, pulling her hair, so one day she ran away with her two kids. She got to Oruro but the police took her back to her husband. Her husband then started proceedings on the grounds that she had abandoned her home, which at the time was illegal. He was very close to winning the proceedings, when they both went to a nearby town and started drinking. When her husband was very drunk, she ran away again, this time getting the train to Villazón and then on to Buenos Aires. She managed to take both her kids with her (Fieldwork notes, 22nd May 2008).

In 2008 I traced the protagonist of this story in La Salada, a large informal market on the outskirts of Buenos Aires, where she owns a stall selling Bolivian food. She was willing to talk and, over a bowl of soup served with a sheep's head, she confirmed her early escape from the mining town. Through migration she managed to free herself from a violent relationship, provide safety for herself, as well as find the means for supporting her children.

Contemporary migration to Spain is also often linked to domestic violence and many in the Cochabamba barrio comment about women not returning to Bolivia because they have violent partners.

Movement and migration were part and parcel of the creation of the mining town, but global dislocations of capital, debt, and structural adjustment policies brought about dramatic changes to one of the fundamental elements on which this place was constructed, that is, work in mining. The price of tin plummeted during the 1980s, which led to internal displacement.

Despite these differences, the way members of this community relate to each other and their own accounts of migration indicate that they share a common identity. Foremost, they relate this identity to their livelihoods, to their identity as miners. Alejandro, a man in his late thirties, who is currently a shopkeeper in Cochabamba and left the mining centre over ten years ago, stated when asked about his occupation: "Well, I am … I am a miner, ex-miner" (Cochabamba, 9th June 2002).

This identity as miners extends to women, whether they have worked in mining or not. The family's dependence on mining as an occupation shapes their identity as miners, or miners' wives. This is not peculiar to this community, and it has also been the case in other communities which rely on mining, and is starkly illustrated in the personal account of the leader of the Housewives' Committee of Siglo XX in Bolivia, Domitila Barrios de Chungara (Barrios de Chungara et al. 1979). A shared history of often extreme poverty and suffering induced by mining activities, such as the common accidents and lives cut short by the 'mining illness' silicosis, bind members of this community together.

The creation of a new 'community': the informal urban settlement

The physical displacement from the mining town to a plot on the outskirts of Cochabamba gave rise to a new set of social relationships and a different articulation with the nation state and the market. The then productive and politically organised miners now became the recent migrants, the marginal urban unemployed, or at best, underemployed: builders and brick makers (Calderon and Rivera 1984). Long-term city dwellers classed them as dangerous, violent drunkards (Goldstein 2005).

The now ex-miners continued to construct their identity in relation to their past as miners and the contrast vis-à-vis peasants was maintained. This becomes evident in their attitudes towards government policies which they see as not advancing the interests of the urban poor. Even if most speak Quechua at home and recognise their indigenous past, many did not agree with Evo Morales' explicitly pro-indigenous rhetoric of redressing centuries of discrimination against indigenous people. This is partly the case because they feel they already found redress through individual means of social mobility. By moving to the city, some gained professional status through further education, sometimes with the money they saved while working abroad. Through time the unemployed ex-miners and apprentice builders became a more varied group of council employees, master builders and professionals (Survey 2008). The mining town now became a more distant place of origin, not so much needed by this more heterogeneous and socially mobile 'community', possibly casting doubts on the extent to which they can be classified as a 'community'.

By moving to Cochabamba, therefore, former miners and their families began building a new community which, over time, diversified to include people who were not originally from the mining community. The majority of those living in the *barrio* were born within the department of Cochabamba and most of these originate from the mining town described in the previous section. About a third of the community are now from other places, including Oruro, Potosí, and La Paz (Survey 2008).

The fact that this is a very young community in terms of its formation is shown in Figure 3.1 below. Only a small minority, less than 5 per cent of the current population, was present at this location before 1990. The majority started arriving to the *barrio* in the early 1990s, when most of the plots were taken up by ex-miners who were displaced from their mining community as a result of the crisis in the mining industry and illness.

Gender relations

Gender relations in Andean peasant societies have traditionally been understood and theorised as being complementary, with women and men having different but complementary roles, which are both valued and essential for the reproduction of rural livelihoods (Allen 2002; Canessa 2012; Isbell 1979; Núñez del Prado Béjar 1975a, 1975b; Skar 1993). However, some disagree with this perspective. As Radcliffe (2015) argues, this idea that Andean gender relations are complementary is a colonial construct, which is often used as a pretext not to intervene into gender inequality. Many in fact have shown that structural gender inequalities persist. The most explicit features of this structural inequality are widespread domestic violence against women (Boesten 2010; Harris 1994; Harvey 1994), the unequal value placed on women and men's activities (Radcliffe 1993; Deere 1977), the lack of formal women's political power (Bourque and Warren 1981), and women's dependence on male labour (Radcliffe 1993).

Despite the fact that many miners in the mining town trace their origins in peasant communities, gender relations in the mining town are markedly

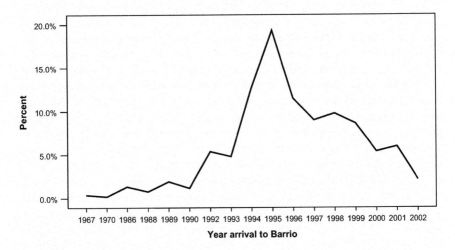

Figure 3.1 Year of arrival to the Barrio

different, with men having a breadwinning role and women being largely responsible for domestic duties and childcare. Gender relations in this community are therefore marked by inequality. To some extent this can be attributed to the heavily masculinised mining occupation and culture which has influenced gender relations in the mining town and, as a consequence, in the *barrio* in Cochabamba. However, as is well known, gender relations do not remain static in migration processes and usually undergo considerable change during the migrants' process of adaptation to their new environment. Despite these changes, gender inequalities persist, albeit in their new, changed form, and are also mediated by the national gender ideology as well as by structural gender inequalities such as gender labour market discrimination.

In the past, women used to form their own household at an early age. In the mining town, it was not uncommon to start cohabiting and have children at the age of fifteen or sixteen. However, girls now study for longer and postpone childbearing and marriage/cohabitation wherever possible. On average, women have eight years of education, while men have ten. These differences are particularly marked for older generations. Women between the ages of 33 and 48 years old have on average seven years of education while men have nine years; and those who are 49 years old and over, have three years and four years respectively. However, for those under 32 years of age, the average number of years in education is the same. The Survey (2008) also showed that the youngest person with a child and living with a partner was sixteen years old, but observations suggest this is now an exception rather than a widespread practice.

Both men and women criticise women engaging in sexual relations outside of an approved relationship, such as cohabitation or marriage. Morality surrounding promiscuity is especially strong and co-exists with a general acceptance that women and men will have sex before marriage or formal cohabitation. These values are put to the test, however, when women become pregnant as a result of having been in such a relationship. Should the couple remain together, the woman's position is not compromised. However, this is not always the outcome of engaging in sexual relations given that it is fairly common for men to abandon their partners once they conceive. This is where gender differences are strongest. Men more readily condone other men not honouring their responsibility for maintaining their partner, whose 'reputation' is then questioned.

Teenage pregnancies are often hidden not only from outside observers but from the community as a whole, although most people know about them. During my various visits to this community, I came across at least a couple of cases where teenage girls had children but these were presented as being their mothers' or their older sisters' children.

Today gender inequalities remain, but it is possible to find some change in power relations between women and men in the neighbourhood. For example, one man in his late thirties presented himself as a 'feminist' when he stated that:

> I admire those people who fight for progress. What I like the most is that men or women fight together. The woman isn't just for the home or for

the bed. No. Women have to be equal to men. Both have to work. That's where you can see progress. (Cochabamba, 20th April 2008)

While this particular example refers to a change in rhetoric, rather than actual practice, there were other cases where the change has gone further. Daniel used to work as a miner and took up small building jobs when they moved to Cochabamba during the early 1990s. He then travelled to Buenos Aires, saving some money to build their family home. Meanwhile, his wife, Melissa, trained as a nurse and started working, initially with temporary contracts and eventually gaining a permanent contract. Daniel continued working in construction, gaining better contracts and hiring his own group of builders for particular projects. They then decided to have another child and when I arrived to Cochabamba it was Daniel who was staying at home to look after the eight-month-old baby while Melissa worked. He explained how the *machismo* so prevalent in the mining town 'disappeared' with the move to Cochabamba, which is partly related to women's labour market participation. When they first moved to Cochabamba he struggled to find a job so his wife began to knit jumpers:

My wife, for example, started knitting and I appreciated her, and my mother, in that they brought water, food, to support their families. [...] So I started valuing those things. Now, for example, I appreciate her. In every home there are always problems. She tells me 'I work', but even with the food, let's say, we're living a moment of matriarchy in this house. I stay at home with the kids waiting for mum to finish work. [...] My mates tease me but I don't pay them any notice because I don't share their views. (Cochabamba 14th May 2008)

Their story is not very representative of the way in which couples arrange and negotiate care work; only two other interviewees, both women, explicitly mentioned a change in gender roles in their household, with their partners taking on more reproductive tasks. Even in this case the role reversal is seen as only a temporary measure, as is highlighted by Daniel when he refers to the fact that they are "living a moment of matriarchy". It is significant that some men now stay at home to look after their children while their partner works. The quote suggests that they are not only willing to talk about it but also do not hide it from their friends. However, the majority of those interviewed did not experience such role reversals and those who did, see these changes as temporary, indicating that there has been a change in gender roles rather than a significant transformation in gender identities.

Feminisation of Bolivian migrations

The Structural Adjustment Programmes implemented in Latin America three decades ago had high social costs. It also brought profound changes to the ways in which women and men relate to one another. Employment programmes implemented to alleviate some of the worst social impacts of the SAPs provided opportunities for women's empowerment and increased decision-making power. At the same time they provoked conflicting and fractured feminine identities (Laurie 1999). More recently, some have suggested that there is an incipient crisis of the patriarchal family, with an increasing number of women refusing to accept patriarchal authority and expectations of traditional gender norms and behaviour (Chant 2009). New social movements brought increased visibility to social groups thus far marginalised, such as the lesbian, gay, bisexual, and transgender community (Lind 2010). There is an increased awareness of the complexity and variety of relationships and family forms, including alternative household structures (Paulson 2010). However, motherhood continues to frame women's identities in their everyday lives as well as in national policies, such as social protection programmes, where some have observed a strengthening of women's identities as mothers (Molyneux 2007).

Migration takes place within this framework of unequal gender relations in which the dominant gender ideology assigns women the main reproductive role as mothers. This has not stopped women from migrating. In fact, the migration of women has increased, both internally and internationally (Menjívar 2005). Despite this fact, however, the migration literature continues to depict women of reproductive age and especially mothers, as having very little autonomy in migration decisions (Chant 1992; De Jong 2000). A study of migration intentions and decisions in Thailand found that "intentions to move were negatively associated with the indicator of dependent child and elderly care-giving responsibilities for rural women" (De Jong 2000: 317–18). For example, after reviewing a number of case studies from different parts of the world, Chant concludes that "even in situations where women are highly mobile [...] men seem to be more mobile still" (1992: 197). She finds that "most women have little choice in determining decisions over their own or others' migration" (Chant 1992: 197), and that motherhood and having key household responsibilities act as a deterrent for women's migration (Chant 1992: 198).

In some cases there is evidence of married women's autonomous migration, such as in Hondagneu-Sotelo's work (1994). In her sample, however, married women openly defy their husbands' wishes with the aim of reuniting their family by following their husbands to the US. Their decisions are autonomous and taken against their husbands' wishes. Moreover, women did not take them with the view of promoting separatedness. Quite to the contrary, their decisions and choices contribute to the reunification of the family, thereby also contributing to the reproduction of a patriarchal institution. As in the case of the Mexican women studied by Curry-Rodríguez (1988, cited in Hondagneu-Sotelo 1994), these married women's strategies aim to re-unite the family and put pressure on their husbands to fulfil their role as household head and breadwinner. As we will see, women from the barrio, also argue that there would be no need for them to migrate, if men provided for them in the way in which they expect them to.

In Asia women have had a longer history of cross-border migration for work, which is often promoted as an explicit development policy by their governments (Piper 2008; Silvey 2004b). However, migrating mothers are sometimes portrayed in a contradictory way as simultaneously economic heroines and bad mothers. In the Philippines, national discourse positions migration as a viable option for single, childless women but not for mothers, who are expected to be by their families' side (Parreñas 2008). While on the one hand encouraging the migration of women, the Filipino state simultaneously enforces domesticity and traditional gender values in which women are seen primarily as the nation's reproducers (Parreñas 2008). To some extent these values are internalised by migrant mothers, who in their stories highlight the suffering they experience as a result of being separated from their children.

Researchers have understood women's migration in Bolivia as being rather passive and saw women as associational migrants until the recent migration to Spain. This is despite the fact that the *process* of feminisation of transborder migration began some 30 years ago (INDEC 1997, see also Table 3.1 below). Thirty years ago Dandler and Medeiros (1988) argued that migration decisions, which they understood as being taken at the household level, favoured men's migrations. They carried out their study within the department of Cochabamba in areas where families engaged predominantly in agriculture. Within their sample, women were largely non-migrants who managed the household by themselves for extended periods while their husbands worked in Argentina. They argued that women preferred to remain in Bolivia to avoid lowering their status by working in Argentina (Dandler and Medeiros 1988). That is, women exercised agency by staying in Bolivia.

A second study, carried out by Balán (1995), largely confirmed these findings, but extended them to include interviews conducted at both ends of the

Table 3.1 Masculinity index (number of men to 100 women) and percentage change for migrants from neighbouring countries living in Argentina, by country of birth, 1980–2010

Masculinity index (number of men to 100 women) and percentage change for migrants from neighbouring countries living in Argentina, by country of birth, 1980–2010					
Country of birth	1980	1991	2001	2010	% Change 1980–2010
Total neighbouring countries	100.40	92.00	86.30	86.55	-13.79
Bolivia	125.40	107.30	101.20	98.68	-21.31
Brazil	85.60	77.30	71.80	72.88	-14.86
Chile	114.70	99.90	91.70	87.08	-24.08
Paraguay	85.60	78.70	73.50	79.72	-6.87
Uruguay	95.20	95.20	92.50	90.80	-4.62

Source: Own elaboration on the basis of INDEC (1997) using data from census 1980, 1991, 2001, and 2010 (INDEC various dates)

migration: Bolivia and Argentina. He concluded that "it seems clear that female migration, in contrast to that of men, is not a response to labour market opportunities" (Balán 1995: 285). Women's role in migration was predominantly associational, where women followed their husbands abroad and did not engage in paid work. Among his sample, some women migrated for economic purposes. However, most of these were generally young, single women who were orphans and needed to seek employment abroad for survival. Therefore, even at the time when an increasing number of married women were joining the labour market because of the imposition of the SAPs, married women and those with children were not engaging in cross-border migration. This is in stark contrast to today.

Current portrayals of the new migration to Spain understand it as *the* event during which women began engaging in trans-border migration in great numbers (Hinojosa 2008a; Román 2009). However, a historical analysis of migration trends shows that the feminisation of migration had already started during the 1980s when more women began migrating to Buenos Aires as a result of increasing demand for women workers (INDEC 1994). Many were independent migrants in the service and manufacturing sectors, particularly the garment sector where Bolivians played an increasingly active role in setting up garment workshops (Bastia 2007). Women migrants accounted for 65 per cent of the total increase in migration from neighbouring countries from 1970 to 1990 (INDEC 1997). Bolivians, together with Chileans, showed the greatest percentage change in the masculinity indices between 1980 and 2010 (see Table 3.1).

This trend was accentuated following Argentina's crisis in 2001, when Bolivians began to migrate to Spain in larger numbers (Ferrufino 2007; Hinojosa 2008a) and where by 2005 over half, 55 per cent, of Bolivian residents were women (INE 2010). The high percentage of women found in Spain did not happen overnight but followed a long process of feminisation of regional migration that provided women with the experience and capital to migrate to Spain. Over half, 57 per cent, of all women interviewed in Spain had previous migration experience in Argentina.

Gendered mobilities from the Cochabamba neighbourhood[2]

Women have always migrated. However, there has been a significant change in what is allowed and acceptable. Men and women from this neighbourhood experienced similar out-migration rates. By 2008, 20 per cent of all women and 24 per cent of all men had migrated at some point in their lives. Migration developed in cycles, increasing during the 1990s in the run up to the Argentinian crisis with a new and larger wave developing in the early 2000s (Figure 3.2).

The new migration to Spain that emerged in the early 2000s was more significant as a destination for women (see Table 3.2). At the time, entry was relatively easy for Bolivians as there was no visa requirement to enter Spain

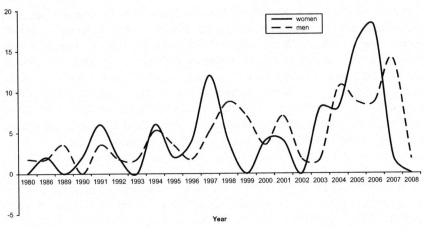

Figure 3.2 Year of first migration, by sex

as tourists. This changed in April 2007 when Spain introduced a tourist visa for Bolivians. It is estimated 70 per cent of all Bolivians in Spain are undocumented (EU 2009).

Many women from the Cochabamba neighbourhood therefore actively participated in transnational migration, often migrating on their own and to a similar degree to men. This was true for young unmarried women as well as for married women and mothers, as was the case for Cochabamba as a whole (Román 2009). It was more common for men to migrate with friends while twice as many women migrated with their husbands. This might suggest 'associational migration' where the woman follows the man, but the qualitative interviews show that few of the women interviewed fit this model (see below).

Mobility is, however, still closely associated with masculinity. Sandra, a single mother, now in her thirties, who migrated to Buenos Aires in 1998,

Table 3.2 Migration destination, by sex

Migration destination, by sex, %		
	Women (N=54)	Men (N=69)
Buenos Aires	50.0	59.4
other Argentina	3.7	4.3
other LA	1.9	7.2
Spain	38.9	14.5
France	0.0	1.4
other Europe	5.6	11.6
USA	0.0	1.4
Total	100	100.0

Source: Survey, Cochabamba, 2008, migrant subsample

despite opposition from her father and brothers, remarked: "I come and go as I please now, as if I were a man" (Fieldwork notes, 28th January 2003). Being single, she is relatively unconstrained in her mobility (in this case, local). She doesn't need to ask for permission to leave the house, as was generally the case for her friends who were married. Her statement supports the idea that mobility and unconstrained movement across space are closely associated with masculine traits.

Women's role as primary caregivers seldom prevents mothers from migrating and there is a wide variety of types of movements and sequencing relating to the migration of mothers. Among those women who had children at the time of one of their migration (not necessarily their first migration), six were single mothers and 24 were married or cohabiting.

There were some instances of strong resistance to the migration of mothers. However, during my stays in the *barrio*, I heard from only one woman who wanted to migrate to Spain and eventually decided not to because of pressure from her husband and father. I did not interview her but recorded our conversation in my fieldwork notes:

> I asked whether she wanted to go to Spain and she said she thought about it and was already getting her documents ready, but her husband didn't want her to go. He said that it would be bad for the family and that they can build their own house slowly. He then agreed to it on condition that she signs a paper saying that she won't have anything to do with her children. She then told her dad and he told her: 'Have you thought about your family? You say you love your kids but you're going to abandon them!' (Fieldwork notes, 8th May 2008)

This seemed to be an exceptional case, given that most mothers I interviewed managed to overcome similar levels of opposition. However, most interviewees, including community leaders and teachers in local schools, and migrants themselves identify the new migration to Spain and the high incidence of women's migration as being closely associated with family disintegration. Marta, a Catholic nun, who teaches in a local school, commented:

> Many families have disintegrated and that's very painful. To achieve a better standard of life, a more dignified subsistence, they migrate, first to Argentina, now to Spain, others to Italy [...] with the aim of bringing back some money, build a house, or pay off their debts. [...] But as a result of that money [...] there is family disintegration. (Cochabamba, 25th April 2008)

Maria, another Catholic nun, went further and linked the absence of parents to the emergence of local youth gangs, drugs, and criminality. Franca, who teaches in a state school, also mentioned this. Another Catholic nun who is active in the provision of social services for the community identified working

mothers more generally as the main culprits, particularly in relation to worsening children's consumption.

A couple, whom I interviewed together and separately at different points in
time, during their joint interview disagreed on this point.

> He: Well, I tell you, in my view it's all about envy. For example, look, she
> has more, she is in Spain, she has this... We are not like this any longer.
> We are envious.
>
> She: Ah, don't tell me that you think that I want to go [to Spain] for
> envy!
>
> He: I am not talking about you. In my view, I am telling her [Tanja]
> this is my opinion. I don't mean everyone. But I think that's the reason
> people go. (Buenos Aires, 7th June 2008)

As others have shown, envy is socially significant among Bolivians (Tapias
and Escandell 2011) and here we see how it is used to criticise women's jus-
tifications for seeking a better life through migration. Marta also commented
on the changes she has seen in the neighbourhood since her return. She is in
her late twenties and went to Argentina in her teenage years and then to
Spain, from where she had returned to look after her elderly mother a few
months before I met her.

> The adolescent isn't an adolescent any longer. They are already youth.
> There are 14-year-old girls who are pregnant. I mean, they escape their
> parents' control. The parents return, but the responsibility or the control
> they should exercise over them has gone. I think that it's because of the
> abandonment that took place at that critical time, just at that stage,
> puberty and adolescence. It's such a critical time, very difficult. And if
> you leave them, well, if they let me free, I'll seize the wings, won't I? I
> think that's what's happened. And there has been a lot of violence.
> Everything decomposed. And that happened only after the migration to
> Spain. (Cochabamba, 16th May 2008)

In contrast to these negative portrayals of the consequences of women's
mobility, many migrant women justify leaving for the benefit of their own
children. This was crystallised in Lucia's account. Lucia's mother took her out
of school at the age of 15 and sent her to Spain to work as a domestic worker.
Her aunts helped look after her. She worked for a couple of years. However,
she had also started seeing a young man and became pregnant. When her
relatives found out, they sent her back to Bolivia. Her daughter was born
only a few months before I interviewed her. Lucia said: "When you have
children, it is as if they had placed themselves in your path and well, time
goes on and you start loving them. I mean, they are always beside you,
accompanying you." However, this love will not stop her leaving her daughter
to return to Spain. When asked about the future, she said: "I was thinking of

going back to Spain. My mum also said, [that it would be good] for my daughter's future. Because the money I brought back from there wasn't that much and my daughter will need more" (Cochabamba, 17th May 2008). Her story illustrates that despite having had little choice about either migrating to Spain or returning to Bolivia, migration plays a central part in her dreams and desires for a better future for herself and her child. Her example also gives an indication of how migrants mobilised different types of social networks, for the different migrations they engaged in, which I discuss in more detail in the next chapter.

Conclusion

Before moving on, it is important to acknowledge that the transnational community was created through cross-border migration, but that internal migration also played a role in this process and helped spur the need for cross-border migration. Gender relations are significant for how this community changed. Despite having a background in an occupation with quite a rigid sexual division of labour, in which men played the role of miners-breadwinners, the migration to Argentina went through a significant process of feminisation. However, unlike previous studies of Bolivian migration to Argentina, women from this community actively engaged in seeking paid work and worked outside the home while abroad.

The regional migration paved the way for women 'leading' the newer migration to Spain post-2001, when Argentina was experiencing significant economic and political stagnation. While this newer migration to Spain is seen as 'the event' in which women led migration, I argue that the regional migration paved the way for women leading the migration to this new destination.

Mobility continues to be associated with masculinity, but there is now greater acceptance of women's migration. The next chapter focuses more specifically on gendered border crossings and the ways in which women and men used social networks.

Notes

1 Parts of this chapter were first published in Tanja Bastia, 2011, Migration as protest: negotiating gender, class and ethnicity in urban Bolivia, *Environment and Planning A*, 43(7): 1514–1529.
2 Parts of this chapter first appeared in Tanja Bastia, 2013, I am going, with or without you: autonomy in Bolivian transnational migrations, *Gender, Place and Culture* 20(2): 160–177.

4　Mobility and social networks

Border crossings

When Marina travelled to Bolivia to collect her four children in 2001, just before the Argentinian crisis, it took her two weeks to cross the border.

> I was working so much, I can't even remember [who was here]. I worked Sundays, I worked Saturdays, I worked … how I can tell you, like a slave. [...] Then I went to collect the kids. I spent two weeks at the border, because they refused me entry so I went at night, I went *por el otro lado,* the other way. I didn't cross at Tartagal, but the other way, with some *colectivo trucho* or irregular busses … I got to Salta. [...] Then I brought the children. (Buenos Aires, 29th May 2008)

Before 2004, some migrants had to resort to crossing *por río*, across the river, or unauthorised borders. This involved dressing up as day visitors and paying somebody to smuggle their luggage across the border, while they went through an unauthorised crossing or borrowed somebody's ID to pretend they were local residents crossing into Argentina for the day. Most often, this was because they did not have enough money to show border officers for *la bolsa*, the bag, as they called the $500 US dollars they had to show to 'prove' that they were tourists and not going to Argentina to work. If they had this amount and presented themselves as tourists, they were given a tourist visa for two weeks, which they then overstayed. If they did not have this amount, they could borrow it but at exorbitant interest rates. One of the people I talked to borrowed $1,500 US dollars for $200 US dollars to cross the border. She was able to keep this money to cross the border and returned it after passing two checkpoint. This is equivalent to a daily interest of 13 per cent. This changed in 2004 with the coming into force of the new migration legislation, which facilitated cross-border mobility.

Both men and women suffered violence at the border. Men were sometimes robbed, while women were often assaulted. Some were physically searched by male officers, who also made comments about their bodies, while others were raped. Monica, Claudia's daughter, first travelled to Buenos Aires in 1998 with her baby daughter:

> I came via Villazón because it was closer to go via Oruro, but they didn't let me in. They told me to go via Yacuiba. So I went to Yacuiba but they also didn't let me in because they were asking for 500 pesos [then $500 US dollars], for the 'travel bag' *'bolsa de viaje'*, as they called it. I didn't have that kind of money. I only had 80 pesos; that was for the whole trip. That's all I had and the only one to be asked to get off the bus was me. Everyone else got through. One of my uncles was also travelling on the same bus. He was detained. They took him in. They didn't do that to me because I was with my baby daughter. Then he told me that they had beaten him up. You have no idea the state he was in at night, all beaten up, full of wounds. He said that the girls that were there, he said that they had been raped, and he said: 'Thank God that they didn't take you too'. (Buenos Aires, 23rd March 2003)

She spent days at the border, with no money, trying to find a way across. In the end, she met a friend who lent her some money. She gave her daughter a sleeping pill so that she would not cry at the border and eventually made it to Buenos Aires.

Crossing the border was also symbolically violent, especially for women who used to wear the *pollera* in Bolivia, the traditional skirt worn by mestizo women (see Figure 4.1). Many felt that they had to change into a skirt and blouse in order to be more presentable and increase their chances of being allowed into Argentina. Claudia had travelled to Argentina in the early 1990s:

> Next day, I woke up, I had a shower, I shined my shoes, and [...] I changed. At that time, I used to wear the *pollera*, that's why I was crying, when I walked, they used to dedicate a song on Mother's Day, 'Mothers of Bolivia, you come, leaving your *pollera* at the border' (*Madres de Bolivia, vienen, dejando su pollera en la frontera*). That's what they used to say and I used to cry. (Buenos Aires, 11th March 2003)

She recalled how her daughters had bought her perfume for the occasion. She bought herself a nice travel bag and new clothes, ready to present herself at the border.

Even those women who did not wear the *pollera* in Bolivia followed a ritual before crossing. I first travelled to Argentina with a Bolivian friend of mine who had lived in Argentina for two years. After arriving to Pocitos, and before crossing the border, we stopped at a hostel. She said we needed to shower before crossing the border. She also took out clean clothes, put on make-up, and changed her regular flat shoes for high-heeled shoes. Migration to Spain involved similar rituals and changes in the ways in which women migrants presented themselves, to convince border officers that they are going

Image 4.1 Woman wearing a *pollera*, Cochabamba

to Spain as tourists. Migrant men did not make similar references to gendered border crossings. Migrant women and men used different social networks to facilitate their migration plans.

Social networks

Social networks play a critical part in migration (Faist 2000; Massey et al. 1993). The role played by social networks in shaping migration decisions and influencing where migrants go to was already evident in earlier studies of migration (Menjívar 2000; Durand and Massey 2006). Over the last few years, there has been a resurgence of interest in migration social networks (Bilecen et al. 2018; Nelson 2015; Ryan and D'Angelo 2018). This latest wave of interest in social networks included a call for a more systematic consideration of the ways in which social networks develop over time and the type of resources that are exchanged within migrants' social networks (Bilecen et al. 2018; Kornienko et al. 2018). Despite early recognition that social networks are not necessarily just 'positive', the tendency remains to focus on social networks as potential sources of support for migrants (Fine 2001). A minority of studies explore the limits of social networks or how they are used to take advantage of co-ethnic relations for economic benefit, for example (Menjívar 2000). This was also the case with Bolivians in Buenos Aires, particularly for those who activated their social networks to find co-nationals to work in their garment workshops (Bastia and Montero Bressán 2018; Montero Bressán and Arcos 2016).

Most of the research on social networks builds on Coleman's definition of social capital, in which he differentiated between bonding, bridging, and linking social capital (Pieterse 2003). Early scholarship recognised that women and men hold different positions within and draw different benefits from social networks (Hondagneu-Sotelo 1994; Menjívar 2000). Some of these studies have shown that: women and men access different social networks (Hondagneu-Sotelo 1994); men and women exchange different resources (Kornienko et al. 2018; Vaa et al. 1989); women derive fewer economic benefits from network membership (Portes and Jensen 1989; Zhou 1992); and that women are often responsible for 'kin work' (Baldassar et al. 2007; di Leonardo 1987).

Methodologically, this latest wave of interest in social networks has highlighted a methodological division in how social networks are studied, with a preference for quantitative studies. This has led some to argue that qualitative methods are undervalued in current social networks analysis (Ryan and D'Angelo 2018).

Women's and men's differential incorporation into social networks has long-term implications for their potential for integration into the destination society as well as for the benefits they can draw from migration (Hagan 1998). Moreover, the gendered dynamics of social networks also have implications for, not only who accesses which social networks and what they can draw from it, but also the geography of social networks. The work by Winters et al. (2001), based upon research on migration from Mexico to the US, shows that: (i) networks play a positive influence on the decision to migrate and on how many people will migrate from a given household; (ii) family and community social networks are substitutes: (iii) once migration becomes more established, households do not need social networks to migrate and that migration takes a dynamic of its own, i.e. community networks substitute for family networks (Winters et al. 2001). This would explain, to some extent, why community networks were more prevalent in the regional migration to Argentina, while family networks became more important in the newer migration to Spain.

Information

Information is the first valuable resource a potential migrant needs in order to make her decision in relation to migration. Whether they migrated alone, with family members, or with friends, all migrants need some form of information before they are able to decide. Friends often provide the first spark of interest in migration. Sandra lived in Buenos Aires for over two years between 1998 and 2001, working in garment workshops and later on, trading clothes at itinerant markets called *ferias*. Her brothers were already in Buenos Aires when she started thinking about migrating, but she recalled that a friend had inspired her to consider migration as a possibility, something she could do:

> I used to see her, the way she used to come, buying the plot, the other plot, building her house. She is also a single mum like me, she still is. She stayed there. She is the one who influenced me the most. (Cochabamba, 25th May 2002)

Despite the important role friends may play in terms of sparking the initial interest in migration, there are some gender differences in accessing information and support at these early stages of migration. Among this particular group of migrants, women often migrate alone and receive little support from their families for the purposes of migration. Participant observation and survey data indicate that men have much better access to friendship-based social networks than women do.

Besides the initial motivation, potential migrants also need specific information in order to decide whether to move or stay. Friends are also critical providers of information about the destination and employment opportunities available to potential migrants. However, it is important to bear in mind that return migrants often distort and embellish the information they provide to potential migrants. Seasoned migrants, and especially men, often feel the need to portray their experience as a successful one. Therefore, they exaggerate their positive experiences while omitting many of the negative ones. This is illustrated by the fact that all of those interviewed were, upon arrival to Buenos Aires, shocked at the sight of the shanty towns where most of them had to live. They were not aware of the high levels of poverty and did not imagine the living

Table 4.1 Social networks: Travel companion, by sex

Social networks: Travel companion, by sex, %

	2002		2008	
Travelled with ...	Women (N=42)	Men (N=54)	Women (N=49)	Men (N=62)
Alone	38.1	33.3	55.1	54.8
Husband or wife	23.8	16.7	20.4	9.7
Friends	4.8	18.5	6.1	14.5
Parents	7.1	7.4	8.2	6.5
Siblings	4.8	7.4	6.1	4.8
Other family	16.7	13	4.1	3.2
Other	4.8	3.7	0.0	6.5
Total	100	100	100	100

Source: Survey, 2002 and 2008, migrant subsample

conditions in the shanty towns because the information they received was well-filtered and selected by the returnees.

Following a positive assessment of all available options in favour of migration, the next step involves the journey. Information collected related to social capital use during the journey abroad indicates that women have lower access to 'bridging social capital' (Pieterse 2003), which is here represented by 'travelling with friends'. In 2002, almost a fifth of all men had migrated with friends while this was the case for less than five per cent of all women. In 2008, most migrants migrated alone, but there was still a significant difference in women being much less likely to travel with friends. Therefore, despite the fact that women 'led' the migration to Spain, men were still better able to recruit their wider, friendship-based social networks than women did.

Initial support and labour market integration

Social networks lower the costs of the migration journey, finding the initial accommodation and first job (Massey 1999; Palloni et al. 2001). Migrants therefore use social networks as a way of finding a place to stay upon arrival in Buenos Aires. Return migrants stated that the extended family was an important source of support when it came to offering initial accommodation (see Table 4.2). In 2008, over a third of all returnees stayed with a member of the extended family. Again, women were less likely to stay with friends.

Qualitative data also indicates that *compadrazgo,* or fictive kin relations established with people at key stages of a person's social development cycle such as birth, baptism, 15th birthday for girls, marriage, and involving often a hierarchical but trusted relationship of reciprocity, are a key part of a person's social capital and used for migration purposes (Rankin 2002). Alejandro, a man in his early thirties who spent a year working in a Buenos Aires garment

Table 4.2 Social networks: Accommodation upon arrival, by sex

Social networks: Accommodation upon arrival, by sex, %				
	2002		2008	
	Women (N=39)	Men (N=51)	Women (N=48)	Men (N=61)
Extended family	35.9	31.4	39.6	37.7
Hostel	28.2	17.6	2.1	3.3
Friends	12.8	19.6	14.6	26.2
Close family	15.4	17.6	43.8	27.9
Workshop	7.7	11.8	0	1.6
Own house/ Other	.0	2.0	0	3.3

Source: Survey, returnees in Cochabamba, Bolivia, 2002 and 2008

workshop explained that he already knew his employers prior to migration: "Yes, almost two months before we met in Oruro. [...] So, we met in Oruro and we got them to baptise our daughter. We were there almost with a guarantee" (Cochabamba, 9th June 2002).

Reference to the 'guarantee' suggests that his job in Buenos Aires was almost secure by virtue of his relationship with the people he met in Oruro. The statement also suggests the possibility that this relationship was purposefully sought after in order to secure the future employment. Another man in his late twenties, explained his plans for migrating to Spain in the following way:

> Look, I have an idea of going to work, going to another country. I have a ... not a relative, she's a *comadre*, we have them here, I mean, the lady who baptised my son, we call them *comadre*. I have a *comadre*, who is from Bolivia, in Spain. She's married to a Spanish man. He had been working here for some two, three years. And she studied pedagogy; she finished and went there with her husband to work. And they are working there. (Cochabamba, 10th June 2002)

His description of the actual practicalities of migrating to Spain are left unexplained precisely because he already has the most important capital that he needs in order to secure his migration plan: the relationship with a *comadre* who will provide the necessary support for his plan to succeed. This passage indicates not only the importance of social capital per se, but the value of the 'linking social capital'. Establishing relationships of *compadrazgo* with people who occupy a higher position up the social ladder, in this particular case the reference to educational level and marriage to a Spanish man, is particularly advantageous. As a strategy, it has been pursued in the Andean countries for centuries, securing strong relationships between the urban and the rural areas, and facilitating usually unequal produce and labour exchanges between them (Gray-Molina et al. 1999; Van Vleet 2008). This particular type of social capital is therefore not new, but it is now used in new ways to facilitate the exchange of resources and services needed for migration to take place.

Migrants generally make use of social capital for advancing their social mobility projects. Interviewees employed in the garment sector generally work for Bolivian or Korean employers. Korean employers demand longer working days and a faster pace but offer better wages in return. Interviewees consider them more trustworthy than Bolivian employers, who sometimes disappear overnight, owing their employees the wages of various months. The qualitative data and participant observations suggest that these preferences give rise to a general pattern where novice garment workers find a placement with a Bolivian employer in order to receive training, which is often unpaid and can last for three months. Following this initial period and once the migrant is sufficiently experienced to run a machine on her own, she can move on to a Korean employer (Bastia and McGrath 2011).

This pattern also suggests that migrants use different types of social capital at different stages of the migration process. Existing bonding social capital is used at the initial stages while bridging or bonding social capital are more important for social mobility purposes later on during the migration process. A number of studies have shown that different types of social capital are useful at different stages of the migration process (Kyle 1999; Janjuha-Jivraj 2003). Gender relations, however, also influence social mobility. Women, having access to fewer occupations, also have fewer opportunities for advancing up the occupational ladder. Migrant men are able to negotiate social mobility, in terms of securing better wages, in a range of occupations but women, who mainly find work in three occupations, find it more difficult to achieve this. In Argentina, nursing was the only occupation that gave migrant women the opportunity to pursue further training and secure their employment as well as their legal migration status prior to 2004. A number of women were also successful traders, but this was seen as a more transitory occupation, albeit one that allows women to accumulate significant amounts of financial capital (Tapias 2015). Migrant men on the other hand have a clearer occupational ladder to follow, which also grants them greater stability. Both men and women are able to become garment workshop owners. The examples that I had access to usually involved couples who bought sewing machines and started employing acquaintances or relatives. This practice often leads to greater class stratification among Bolivians, in which garment workshop owners use their own social networks to find employees, who often end up working in exploitative working conditions (Montero Bressán and Arcos 2016).

Collective action

The shanty town, Las Achiras, where a number of my respondents live is adjacent to the Central Market in Villa Celina, province of La Matanza, Buenos Aires. The military government built it as a transitory neighbourhood during the 1970s. These neighbourhoods were to house the shanty town dwellers expelled by the government from the Federal Capital as part of its shanty town eradication programme, but only temporarily. Government policies later changed and those housed in the temporary accommodation were allowed to remain. This initial nucleus of houses slowly expanded as demand for accommodation close to the Capital increased and the area became semi-urbanised.

The government provided basic services like water and electricity for the government-built houses but not for those built later on. Doña Claudia, who was one of the first women to have migrated to Buenos Aires from the mining town, settled in Las Achiras during the 1990s. During the interview, she described in detail the beginning of the urbanisation of that particular area. The majority of those buying the land were Argentinians, but there were also some Bolivians. Doña Claudia described the way Bolivians organised to bring water and electricity to the community:

> Do you know how we brought this [water]? We used to go to dig at night, over there, to bring that water. [...] We [Bolivians] put down our quota for the tubes, the keys, to break [the tube]. So the other one who already knew a bit about the process, he cut the thick tube that was passing over there. He cut it and there was no accident. Everything was ready and we soon had water. For the electricity, it was the same. We used to use candles for a long time down there. Now we already have electricity. The Argentinians didn't want to chip in one little cent to bring the posts and they were the majority over there. There weren't many Bolivians. But we had to make the contributions because the Argentinians are always broke. (Buenos Aires, 11th March 2003)

In another neighbourhood in Villa Lugano-Villa 20 in the city of Buenos Aires, there were other attempts to organise this community of ex-miners for wider-reaching purposes. Javier is a university-educated social worker in his early thirties who has lived in Buenos Aires for six years. He was born in the mining centre, but moved to Oruro to study at high school and later to Cochabamba to university. During the interview, he described one initiative that aimed at organising the whole community of ex-miners for the benefit of its members:

> We started because, you would have seen, there are many people from the mining centre here. It's more, I came to live here [Villa Lugano] because I found so many people from the mining centre and well, at some point we tried to create a society, let's say a group that would have functions besides sporting activities, such as projects with medical institutions to help out. (Buenos Aires, 28th March 2003)

As with migrant organisations elsewhere, the ways in which Bolivians organise is also highly gendered (Strunk 2014). In Buenos Aires, migrant men tend to get together to organise sporting activities at the Parque Indoamericano on the weekend (see Image 4.2 and Image 4.3). However, the leadership of the organisation that Javier was active in, which was made up entirely of men, decided that the accommodation needs of newly-arrived migrants were of foremost importance, followed by their need for medical insurance. Unfortunately, the crisis of December 2001 prompted many of those involved to return to Bolivia and the organisation dissolved, but there is a lesson to be learnt in this process. Javier described the main reasons behind the potential for this organisation:

> We couldn't advance much, but it was really helpful to work in a group here in the neighbourhood because there are lots of very positive things from the people here. I am referring to the people from the mining town: between us, there is a lot of closeness and support[1]. Let's say that I need to build the roof. I know who to go and look for, and I know that they will be able to help me and I know that they will not charge me. And those who can come to call on me they also know who can come, all between us from the mining town. They come and tell me 'Look, come and help me because you also need [help]'. (Buenos Aires, 28th March 2003)

Both of these examples illustrate the different ways in which men and women participate in collective action. Men are more likely to lead formal

Image 4.2 Parque Indoamericano, Sunday get-together, Buenos Aires

Image 4.3 Parque Indoamericano, Sunday football matches, Buenos Aires

organisations, whether these are based on national identity or place of birth, as in Javier's case. Women, on the other hand, are much more likely to participate in informal activities that aim to improve collective well-being, as in the case of Claudia. Women in this case tend to remain within horizontal social networks while men seem to be more strategic in terms of using vertical social networks that cut across nationality ties (see also Bastia, 2018).

When social networks fail

Clearly social networks are not just 'positive'. There is a tendency to consider social capital as something that is desirable (Fine 1999). However, undesirable social organisations, such as gangs and mafias, also rely on high levels of social capital. While I did not come across such examples, what seems clear is that women and men had differential access to social networks. This difference crystallised when I observed two instances of a man and a woman in crisis, but with very different outcomes in terms of how they were able to access support from their 'community'.

The previous section discussed two instances where migrants used their bridging social capital and bounded solidarity to overcome some of the problems they have encountered in Buenos Aires. The first example depended on a wider collective identity based on national origin, Bolivians, while the second one was limited to those related to the mining town and as such, a subgroup of the national group. These examples showed that migrants use different identities for the purposes of capitalising upon their social relations. Javier, the social worker who described the attempts at organising the community, is a striking example of the way origin-based identity mediates migration-related decisions and provides a source of support, therefore effectively providing the basis for building social capital. He is also conscious of the fact that a strong bond of reciprocity and common origin unite this 'community', as his following statement shows.

> I, to tell the truth, I had lost touch with the people from the mining town because I left when I was small and I never returned. [...] You know, I was really from the mining town, so, in that way I got in touch. [...] Yes, because the people [from the mining centre] are always in contact. That is good, that is what can be salvaged from Bolivia. Let's say, we have never lost our unity. (Buenos Aires, 28th March 2003)

His example is interesting because he has not followed the typical trajectory of other members of his community. He never settled in the neighbourhood in Cochabamba and lost contact with this community throughout his high school and university years. However, in his life account he talks of 'returning' to the neighbourhood in Cochabamba, despite never having lived there. He re-established his relationship with this community as part of his migration strategies.

It is clear so far that the shared identity and bounded solidarity is often the basis on which migrants are able to use social capital in order to gain resources or services from members of their own kin networks or the community as a whole. Sometimes this extends to the wider national group, as was illustrated with the collective action undertaken to secure water and electricity in a shanty town. However, social capital theorists also caution against over-optimistic views of social capital and its consequences (Portes and Landolt 2000; Rankin 2002).

Migrants' accounts provide ample examples of cases where fellow countrymen and women – *paisanos* – use what is supposed to be their bounded solidarity for personal financial gain. There is a certain longitudinal element in these relationships. In some instances, better-established migrants with longer residency in Argentina take advantage of the newly arrived migrants. This statement needs to be qualified as this is not an outsider's view of the situation[2]. Most migrants are willing to undertake long working hours for lower-than-average wages because they still compare positively with the options they have available in Bolivia. Therefore, it is not the exploitation per se which migrants find 'exploitative' as such, but the fact that their fellow *paisanos* take advantage of their position in Buenos Aires.

Besides its 'negative' uses, there is a further problem with the often over-optimistic view of social capital. Social capital is a useful resource for job hunting and for obtaining accommodation. However, the real test is whether social capital can provide support in times of crisis and if so, to whom (Silvey and Elmhirst 2003). Migrants' accounts provide evidence to both accounts and the analysis of two particular events will shed light on the internal dynamics of social capital within this community.

The first example involves the death of a young woman who was then living in Buenos Aires. The community held her as an example of success. Her partner left her when she became pregnant. She brought up her daughter on her own while working and studying in Buenos Aires where she had successfully trained and qualified as a nurse. She also found a stable, legal job at a hospital/nursery home. Her loss breached the physical distance that separated those living in Buenos Aires from those in the neighbourhood in Cochabamba. As the news of her death travelled from house to house and people began mourning, they also started thinking of ways of getting her father, then in Cochabamba, over to Buenos Aires for the funeral. He took a plane the following day and was able to see his daughter off.

The second event is related to Maria, a single mother I interviewed in 2003. She had been living in her brother's house for a while and had had her second baby only a month before this incident took place. She was visiting her friends one night and decided against taking the risk of walking home in the early hours of the morning because of the high incidence of crime. The following day she was severely beaten by one of her brothers. Other family members were also present in the house at the time. Nobody intervened. She moved out of the house with her baby and having nowhere else to go, she asked an Argentinian neighbour for help. Here she found shelter and support. She decided against pressing charges, despite the severity of the beating, because of the consequences of such action; she would be ostracised by her family, including her mother.

Comparing these two events provides some evidence to support the arguments that social networks are not equally available to all those who belong to a specific 'group' and that support and help in times of crisis is highly gendered. In the first instance, a senior male member of this community of ex-miners was able to use his social networks to borrow money and fly from Bolivia to Buenos Aires to attend his daughter's funeral. In the second example, a younger female member of this community was unable to find emotional and practical support within her community after her brother abused her. Despite the fact that the first example involved a monetary transaction and arguably a higher risk factor (the money could not be returned) while the second one involved emotional support and temporary shelter, he was able to draw on his broader social networks within the community but she was not. This shows that there are significant internal inequalities based on unequal power relations within the community that influence people's ability to capitalise on their social relations. Therefore, despite being able to migrate and often doing so autonomously, women migrants may find themselves isolated in times of need.

Sequencing

Terminology plays an important part in how we perceive migration and the processes associated with it. Much of the recent literature talks about the 'feminisation' of migration, but without unpacking what this really means. At the superficial level, the feminisation of migration refers to a numerical increase in the number of women migrating, in relation to the number of men. As Donato and Gabaccia (2015) explain, the terminology used is often inaccurate even in the numerical sense. However, I am also interested in what this means for the power relations implicated in the process of migration, including decisions about migration.

Women migrated in a number of modalities. Of the 24 women who had children and a partner at the time of one of their migrations: six migrated with their husbands, six followed their husbands, four migrated first and were later joined by their husbands and eight migrated on their own. One of the latter migrated to Spain on her own, applied unsuccessfully for family reunification on multiple occasions, then returned to Bolivia and went back to Spain with her husband. There is therefore no clear pattern related to the sequencing of migration.

The terminology related to 'feminisation' gives the impression of women having a greater say in their migration journeys and the decisions surrounding migration. 'Male-led' migration, widely used in the literature, implies that the journey and the decisions to migrate have been taken by men. 'Women-led' on the other hand, implies that women had a greater say in their decisions and their journeys. The term used to describe the act of a couple migrating together, 'joint migration', gives the impression that the decision was also shared. However, these assumptions obscure a number of important elements, which are related to how many times migrants undertake their migration journeys, the temporality of migration, the decision-making process, as well as autonomy of migration.

Multiple migration journeys

The language related to sequencing tacitly implies that there is a single journey to consider. Although some migrants do migrate only once, this is by no means always the case. Many migrants engage in multiple journeys, back and forth between two places, but often also between different places. This means that the same person can also engage in more than one sequencing in relation to her journeys, those of her partner, and the ones that they might take jointly.

Ana's experience is interesting because it illustrates how even within the same relationship, women can practise different sequencing, but also have very different experiences of autonomy within the migration decision-making processes. During her first migration, she was a 'follower', i.e. she followed her husband to Buenos Aires to help him pay off the debts he incurred while putting in practice his dream of owning a transport business. Despite the fact that she supported him during the first few months, migration and earning a living was clearly his responsibility. She 'helped' when it was necessary and left her job to care for her son as soon as this was needed and they could afford it. They returned to Bolivia after over ten years of living in Buenos Aires, but she grew unhappy with the relationship. They lived close to her mother-in-law and his transport business was still not delivering the desired profits. They were unable to finish the house and live 'as people'. She also felt unloved and suspected that her husband had cheated on her. Therefore, she borrowed some money from her mother, left her daughter at her sister's house, and went to Spain. A few months later, he followed and they then got back together (San Fernando, 13th May 2009). Here we have an example of a woman who was a 'follower' during her first migration but then 'led' the second migration, in which her husband followed her to Spain.

Temporality

When discussing migration sequencing and decision-making, there are two aspects related to temporality that are often obscured. First, the fact that a woman-led migration, for example, can begin as a 'woman-only migration', when the woman migrates on her own, but when she is later joined by her partner, this becomes a 'woman-led migration', as in the case of Marina. Therefore, it makes sense to talk about sequencing, but only when this refers to a specific time frame e.g. the characteristics of a specific migration journey within the first year of being abroad. To return to Marina's example, she travelled from Bolivia to Argentina on her own, but her husband decided to join her eight months later, when he realised that she was not coming back. Is this a woman-only migration or a woman-led migration? To what extent are these labels useful when describing this migration journey? If we take a six-month temporal frame, hers would be classed as a 'woman-only migration', but if we take a one-year temporal frame, it would be classed as a 'woman-led migration'. Clearly, the timing of the interview would also give a different impression of what was happening. If interviewed six months after her journey, hers would be classed as a 'woman who left her husband' journey, while one year on, they would have been reunited in Buenos Aires.

Second, the decision related to migration might be taken and acted upon almost instantly. Raul, for example, who had already been in Buenos Aires during the 1990s, but returned to Bolivia because he did not like the work nor the lifestyle there, one day had an argument with his brother and decided there and then to return to Buenos Aires. He left the following day and eventually settled in Buenos Aires, working as a bus driver (13th February 2002). This decision, taken in the spur of the moment, led to a 'permanent' migration to Buenos Aires, or one as permanent as can be. Having migrated during the 1990s, Raul was still in Buenos Aires in 2009.

For others, the decision to migrate might involve lengthy negotiations that can take the best part of a year. Rosa had travelled to Argentina at the age of

14 with her mother and some of her sisters, and studied for a few years. She also worked in the garment sector and in elderly care. She then went back to Bolivia, finished high school and entered university. "I didn't want to get married" but while she was studying, she became pregnant and married her partner. However, unlike many of the other women, who had to drop their studies once they married or started cohabiting, Rosa managed to finish her degree, with the support of her mother-in-law who helped her with childcare. For her, migration was an opportunity because she wanted to "have something" and tried to convince her husband to go to Spain:

> At the beginning he didn't want to come because his mother wouldn't let him. So I stopped insisting but my idea was to finish my degree once and for all, and then 'I'm going, with or without you'. (San Fernando, 27th July 2009)

The process of convincing her husband took about a year, which illustrates how decision-making is not an event but a process, one which often involves a considerable amount of time. When her daughter turned one, Rosa's husband asked her to borrow some money so they could travel to Spain. He went first and she followed a few months later with her daughter. She found a job in elderly care, working nights while he worked in a campsite. At one point she felt that they were hardly seeing each other, so she halved her job with her sister, so she could have more time for her family. She then worked alternate nights and had a cleaning day job, paid by the hour (San Fernando, 27th July 2009). So in this case, he travelled first but on her insistence and the process of making the decision took a year.

Joint migration: not necessarily a joint decision

Women also exercised a relatively high level of agency in cases where they migrated jointly with their husbands. There were six cases of joint migration. In two cases it was the woman who decided that they should go, but through a lengthy process of negotiation, as discussed above. Other examples give indications that even in cases where both partners migrated together, the decision was mostly taken by the woman. Leandra, for example, described how her husband's mother was resentful of her because she was the one who wanted to go to Argentina. When asked whether he wanted to go to Argentina, he replied: "I was young. I cannot tell you whether I wanted to go or whether I didn't want to go" (San Fernando, 29th July 2009), an ambiguity that he resists clarifying.

Sonia followed her mother to Buenos Aires when she turned 18. She wanted to work so she could return to Bolivia and study. During one of her trips back to Bolivia she met her husband and convinced him to go to Buenos Aires, something for which his mother resented her to this day. She had her first child shortly after and could not fulfil her desire to study. She relates this wish to her own experience, growing up in a family with domestic violence: "I would like to study. I would like to defend women who are being maltreated" (San Fernando, 29th July 2009). They returned to Bolivia just shortly before the crisis, opening a restaurant, but she did not work. By 2004, they had no money and had incurred a debt to build their own house. She was considering going to Spain, where her mother and sister were at the time, but she could not face leaving her 8-month-old baby and her older son. She describes how her sister's children call their grandmother "mum" and felt that "I don't know what I could do without my children. I would go mad" (San Fernando, 29th July 2009). So, she convinced her husband that they all had to go to Spain together (see also Chapter Six on care).

Given the nature of the labour market at the time of the Bolivian migration to Spain, a joint migration does not imply that both men and women worked while abroad. There were many cases where women had to support their husbands while they were abroad, because they had been made redundant in the financial crisis or struggled to find work in the first place. Patricia travelled to Spain with her husband in June 2005, borrowing $2,200 for each flight from the bank. She had a debt with the bank because she had bought a car, hoping to work it as a taxi. She and her husband had also taken out a loan for the house, for which her mother was the guarantor. But the car wasn't giving them a lot of money and they were unable to repay the loan. Her husband was working with the car and then as a builder, earning 45Bs. per day (about $6.5 US dollars). The bank was also harassing her mother, as the guarantor, so they decided to go to Spain, borrowing $4,400 US dollars for the flights, leaving their son with her mother. On their first attempt, only her husband got through. She was deported together with another 80 migrants "because they could see I was going to work". She tried again one month later and got through "if I hadn't gone through I would have killed myself". By then, they owed the bank around $6,600 for the flights, plus the car and the house. She first found work bathing an older man, three or four days a week, two hours each day with a salary of seven euros per hour. She did not stay long. Her next job involved cleaning an elderly woman's feet. After about three months, she found work looking after a woman with Alzheimer's. She worked there for over a year, six days a week, 24 hours a day and only having Sundays free. The woman used to scream from the window, saying that she does not want her in her house. She used to beat her as well. Meanwhile, her husband only worked from time to time, when he found jobs on a daily basis in construction. Therefore, she had to support him as well throughout their stay in Spain (Cochabamba, 20th May 2008).

Autonomy: Women-only and women-led migration

Lucia was 15 when she was sent to Spain by her mother to repay a debt (see Chapter Six on care). She worked there for three years, before coming back to Bolivia, eight months pregnant. In her interview, she expresses sadness about being taking out of school and having to work to repay her mother's debts. However, she was not given much choice in the matter. The low level of agency displayed in Lucia's story is rather exceptional, since most women showed great autonomy and decision-making power in their descriptions of the decision to migrate.

One would expect women-led migration to be mostly autonomous in the sense that women who engaged in this type of migration would also take the decision to migrate by themselves or at least, would play an important role in making the decision. This is the case for most of the women who migrated on their own but by no means all of them, as seen above with Lucia.

Fernanda had never been abroad when she decided to go to Spain in 2004. Her husband was then working in another city. She arranged all her documents, called her mother to come and look after her children and then informed her husband only days before she left:

> He was surprised. He didn't say anything. I told him 'I'm going to go. I need to get money for the house and you are going to give me so much [for the trip]'. He gave me more than I needed so I lent some to my friend [they travelled together]. He didn't know. He still doesn't know. (Cochabamba, 29th April 2008)

Fernanda's experience was quite different from Marina's, who was living in Cochabamba with her husband and her three children. Her husband was then working as a building contractor, earning relatively well compared to his neighbours. They had incurred some debts because they had been building their own house. She decided to migrate despite opposition from her husband and *compadres,* or fictive kin.

> We tried to get on with things a little bit; we tried improving our situation. You want to achieve something and then, my daughter was one year old and they [neighbours] were talking about Argentina. They talked to me about it and told me that one can earn well. So I said: 'I'm going' because I was also sorry to see my husband working in full sun as a builder. I felt sorry for him. He was the only one who was working at the time. In Cochabamba, I tried to help him. Small things, but I wanted to help him a bit more. Then there was a neighbour who was encouraging me to go [but then] she told me she didn't have enough money so she couldn't travel … . But I said: 'I'm going.' So I came without knowing anyone … I came on my own. (Buenos Aires, 7th February 2003)

Her husband Carlos confirmed this account: "Marina said: 'Let's go. I'm going.' To tell you the truth, I don't know, I didn't want her to go because my *compadre* said: 'No, don't let her go' … But she came to Buenos Aires to work" (Buenos Aires, 2nd March 2003).

Marina's migration was autonomous in the sense that she decided to seek work abroad despite opposition from her husband and their *compadres*. He explained that they owed money, but the debt was not substantial, about $200, at the time equivalent to a labourer's two months' wages. The level of income earned by Marina's husband was substantially higher. In fact, they had been able to build their house in a year. Immediate dispossession and unsustainable debts were not the main drivers of her decision to leave. She decided to migrate because the choice of staying did not include the possibility for upward social mobility, to see her children study and become professionals, to have something more than just 'getting by'. Clearly, her decision was embedded within everyday social relations and the aspirations for social mobility expected for those who can seek work abroad. This decision is also understandable once we place it within the broader context of her life story: the deprivation she experienced in her own childhood and the social exclusion miners suffered when they migrated to cities. However, the fact that relatively speaking, her household was not economically deprived coupled with the fact that she took the decision in opposition to her husband indicates that she exercised a high level of agency and acted autonomously, though not necessarily in self-interest.

We can appreciate this high level of autonomy both in terms of setting the social mobility objectives her family should aspire to and having control over her own geographical mobility. In a later interview, she explained that she would not have migrated, had her husband taken responsibility for earning enough money for their family (see Chapter Five on work). This is similar to the justification provided by married women interviewed by Hondagneu-Sotelo (1994), except that in her case, women were migrating to join their husbands, while in this case, women are initiating migration trajectories.

Interviews with men whose wives migrated to Spain also indicate that women often took the decision by themselves. It is interesting that some men related the fact that they let their partners leave to having a progressive predisposition. Jacinto, for example, met his wife in Argentina during the 1990s. They had a child together and then returned to Bolivia in 2002 because of the *corralito* following the 2001 December crisis. They came back together "but she had other plans" he says "of going to another country". The new migration to Spain was in full swing at the time. He said he did not want to be like other men he used to see in the neighbourhood, who had four or five children, working as builders, drinking and going nowhere. He wanted to study. However, they disagreed. She said he would never pass his exams. "Well, you want to go, go then, so that you won't say that I am *machista*". He wanted to build a house, but she said no. Therefore, he gave her the money and told her to go. "She left ... [Pause] She came back after three years. [Repeats, in a lower tone]. She came back after three years. Then she took our son. 'Are you coming?' 'No', I said. 'I will finish studying'" (Cochabamba, 20th April 2008). She ended up staying in Spain and he finished his degree. However, they separated and today they both have different partners.

Most women initiated their own migration because they were unhappy with their partners, either because they were not collaborating, they were not present, or their earnings were insufficient to fulfil their ambitions. Women were unhappy when their partners were not fulfilling their 'role' of economic provider, but they also complained when they felt that they were not emotionally close with their partners. Interestingly, some associated 'emotional closeness' as a characteristic of intimate relationships with Spanish partners (see Chapter Seven on intimacy). Given the opportunities available to women abroad, they took the initiative and migrated. This was the case with: Valentina, who left her children with her husband only to find out that he had left them on their own to go to the Chapare; Marina, whose conflicting ideals of gender roles and responsibilities were described at the beginning of this section; Claudia, who was in Buenos Aires with her husband but he hardly featured in her testimony; Josefa, who wanted to build a brick house and needed money for the upgrade; Catalina, who wanted a little house and for her children to study but prices kept increasing and they could not make ends meet; Erika, whose husband gave up an office job to join her in Spain and was then pressuring her to go back, but they stayed in Spain; and Elvira, who wanted her children to have a better education. They are all women who had ambitions for their families and their children. They wanted to either improve their house or make sure they had sufficient money to pay for their children's education. In a number of cases women also wanted to escape domestic violence or unhappy relationships.

In three cases, the women decided to follow their partners on their own accord. Two of these cases involved men who went to Argentina and then stopped communicating or sending money to their wives in Bolivia. The wives then decided to travel to Argentina and join their husbands, in a similar way to the Mexican women in Hondagneu-Sotelo's (1994) sample.

Ana's husband had gone to Buenos Aires with some friends and she followed shortly after. However, she found that she had to support her husband because there were no jobs for him:

> I was working, because at that time there was more work for women and my husband couldn't find a job, I was the only one working, we were renting a room before buying the house, I worked for that first year. Then my sister arrived and my brother-in-law so we had to buy the house there, we had to borrow money and it took us almost a year to pay it back. We owed that money to my husband's cousin. (Cochabamba, 28th May 2002)

They returned to Bolivia after the 2001 crisis and with their savings, they managed to expand their house and pay some of their debt off. However, they still needed more money to finish their house and to repay the debt. Things were not going very well in their relationship either so in 2005, she decided to go to Spain, on her own.

Ana's story shows that within a lifetime, women experience different migration modalities: male-led, joint, then women-led with migration being undertaken for different reasons, sometimes they are economic, at other times these may be more closely related to unhappiness in their intimate relations. However, the point that I wish to highlight here, is that the modality of migration does not indicate that the person who 'leads' the migration, i.e. who migrates first, has the most say in the decision to migrate. In fact, some, who migrated jointly with their husbands, spoke about their migration experience in the singular form.

Josefina, for example, followed her husband but he hardly features in her story. Josefina was in her early thirties, married and with four children at the time of the interview. She migrated three months after her husband, leaving their children in Cochabamba. In her life story account, Josefina talks of her work and the decisions made while in Buenos Aires quite independently of her husband, filling information about what her husband was doing only when prompted by my questions. In fact, she begins her life story by talking of her experience in Argentina in the singular tense, as if she had migrated on her own: "I was born in Cochabamba. I went to Buenos Aires to get my house, because I didn't have a house. I went, stayed there for three years and after three years I came back. Now I have my house" (Cochabamba, 8th June 2002). Josefina mentions her husband only five minutes into the interview, when asked to describe the first trip to Buenos Aires and after having talked about her leaving her children in Bolivia and her furniture at her nephew's house in Buenos Aires. She explains that her husband was not earning well: "I was working, but I was also sending money for my children and we had to pay rent, all that, it wasn't enough". Towards the end of the interview, she highlights her own achievements, speaking again in singular: "Yes, the six months I was there, it didn't go so well, but the three years went well. Yes, I could save some money, I bought my house. I had a debt with the money I borrowed to go there [to Argentina], I paid my debt, all of it" (Cochabamba, 8th June 2002).

Despite high levels of autonomy, for many of the women interviewed, autonomy is not the objective of migration. Rather, the high levels of autonomy are a means to an end, aimed towards material achievements and upward social mobility, which are discussed in more detail in the next section. The stories also show that these projects are usually aimed at increasing collective family well-being rather than meeting their own individual needs and interests, although the distinction between these two is not as clear-cut – and in many ways, the two are related.

The women who 'followed' their husbands showed different degrees of autonomy in the process of migration. 'Male-led' migration does not necessarily mean that men take the lead. Sometimes women decide that they should migrate; or both partners make a joint decision that they will migrate and men go first.

Motivation: predominance of the material?

Money, work, and material success were the most often stated objectives of migration. Most migrants stated that they migrated for work. As in Thailand

(De Jong 2000), there was very little if any difference between women's and men's stated reasons for migration. The last survey in 2008 indicates that 86 per cent of women and 87 per cent of men from *barrio* stated that they worked while abroad, despite the fact that Bolivian women in Argentina are generally associated with low levels of labour market participation compared to other migrant nationalities (Zunino 1997). However, there is some difference in their initial intention, given that 89 per cent of women, compared to 84 per cent of men, migrated for work. That is, men had a slightly higher likelihood of migrating for reasons other than work, but then in practice a similar percentage of both men and women worked while abroad.

While there was very little divergence between women's and men's motivations in the quantitative data, the qualitative life story interviews, in which migrants were better able to elaborate their own stories of migration, showed that men were much more likely to state that the initial impetus that led them to their decision to migrate was for reasons other than work. In fact, 87.2 per cent of all women interviewed explicitly stated that they migrated for work while this was the case for only 70 per cent of all men migrants interviewed – not an insignificant percentage, but lower than that of women. Some men migrated because they were curious; others because they had a fight with somebody or because a friend had returned from Buenos Aires and encouraged them to migrate with them. One man migrated because he wanted to join his wife.

Women, on the other hand, stressed the importance of work and material attainment in motivating them to seek work in Argentina and Spain. For many women, migration is an avenue for acquiring capital which would otherwise be impossible to save had they stayed in Bolivia, as in the case of Josefina, discussed in the previous section. Material goods satisfy basic needs, but, in many cases, they are also used to signal social mobility. Diana, who was 24 at the time of the interview, stated that: "Here in Cochabamba, I was well. I was with my baby, with my baby daughter; I was well. But then, to have more things for my children, I went to Buenos Aires" (Cochabamba,

Table 4.3 Main reason for migration, by sex

Main reason for migration, by sex, %		
	Women (n=47)	Men (N=30)
Work	87.2	70.0
Curiosity	0.0	16.7
Find wife/husband	2.1	3.3
Other	4.3	3.3
Unspecified	6.4	6.7
Total	100.0	100.0

Source: Selected life story interviews, Cochabamba, Buenos Aires and various cities in Spain, 2002–2009

10th June 2002). Similar motivations were recorded in Thailand, where women migrate as a means of achieving modernity (Mills 1997) and in Vietnam, where women migrate to match the increasing expectations of consumption goods in farming areas (Resurreccion and Khanh 2007).

For Ana, her objective was to finish building her house and live "like people, I don't expect more, to always have luxuries. No. Just to live like people, because I have lived in poverty. I don't have a father or a mother. I never had a home. At least I want to live like people" (Cochabamba, 28th May 2002). This statement suggests that it is only through international migration and work in foreign labour markets that these recently arrived marginal urban populations will be accepted as part of the city and overcome the historic discrimination they have experienced.

However, it would be a mistake to take these answers solely at face value. Even when economics is the driving force for mobility, this does not mean that we can or indeed should separate the economic and the cultural dimensions of migration (Rigg 2007). When men migrate 'for work' they are acting out their breadwinning role abroad, a relatively simple displacement of the main role assigned to them in this particular cultural context. Yet for women, migrating 'for work' implies quite a radical break with their expected roles as mothers, daughters, and wives within a cultural context which prefers reproductive activities to be carried out in close proximity to the sites of production. Migration is to some extent challenging normative assumptions about women's and men's roles, though this does not necessarily mean that this process creates new gender identities.

Although couched within a discourse of need and economic deprivation, at least four of the women's life stories indicate that their migration was linked to a project of *separatedness*, i.e. deliberately choosing a separate life, one that is independent of, or at least to some extent disconnected from, their partners. Doña Claudia began migrating during the 1980s, when she left the mining town. Her story is rather ambiguous, mainly because her husband is hardly present in it. Some of her daughters stated: "my parents they didn't live well" and went on to describe how their father often beat their mother. Their father also did not provide for the family financially and doña Claudia worked, first in Bolivia as a miner and trader, and then she travelled to Buenos Aires where she worked in various garment workshops and as a cleaner to support her five children.

Sometimes these projects of *separatedness* were directly linked to domestic violence (see also Chapter Seven). Domestic violence is very common in the neighbourhood. Only some cases, those deemed 'extreme', are condemned. Doña Claudia told me the story of another woman, a neighbour, who ran away from the mining town to escape her violent husband. "She was older than me, but I used to help her hide because her husband was very bad. That's what she ran away from, from there, from Bolivia. I would open the window and hide her in my sister's house. That's how I met her" (Buenos Aires, 11th March 2003).

When I returned to Buenos Aires five years later, I found the protagonist of this story, doña Paolina, who confirmed her escape from her violent husband. She did not give too many details of the day she ran away, but chose to highlight her achievements: a successful business in the biggest informal market in Buenos Aires, where she owns various stalls; a son and a daughter, both with good jobs; and a partner, who helps her with the business and keeps the burglars at bay, who was happily dozing off while we talked over piles of plates that needed washing (Buenos Aires, 2nd June 2009).

Ana and Fernanda had also effectively used migration to separate from their husbands, albeit temporarily. Fernanda, whose story was discussed above, left her husband to go to Spain, but then returned because she missed her children. She does not state this explicitly, but the way she describes her migration gives the impression that she had left her husband. She returns to this point later on in her story:

> Being separated has united us. My husband is now good. He doesn't say anything. He accepts everything I say. He doesn't say anything. Before, he used to argue with me. Maybe it is because he thinks that I'm going to leave him again, I don't know. He understands more, maybe because of his age. He understands me better and I'm more hysterical. [...] Before it was the opposite, I was good and he was uuuh, terrible! (Cochabamba, 29th April 2008)

Her language switches from talking about being physically separated at the beginning of this paragraph, and then moves on to refer to her 'leaving him', suggesting that she had used the physical separation to end or at least temporarily suspend their relationship. This was similar to Ana's experience, mentioned above, who grew unhappy with the relationship she had with her husband when they returned from Argentina, so she went to Spain (San Fernando, 13th May 2009).

The stories in this section illustrate that for most women, their dreams and desires include financial security as well as hopes for upward social mobility for their children. For most, migration is a means of achieving greater collective family well-being, rather than being linked to the pursuit of *separatedness* and individual self-interests. For a small number of women, migration did indeed include the desire to separate from their partners. However, in most of these cases, they had not only grown disillusioned with their relationship, but had been often severely abused by their husbands. For others, such as doña Claudia, whose husband drank excessively, migration represented her and her children's means of survival.

Conclusion

This chapter has shown that the modalities of how men and women migrate, whether the migration is 'male-led' or 'women-led' say little about the actual decision behind migration or the autonomy that women may have in these negotiations. Some male-led migrations were undertaken because of women's decisions to send their husbands first to Spain or Argentina. Some women took the lead in their migration because their husbands were absent or did not fulfil what they deemed as basic responsibilities. Others did not live up to their ambitions of having a brick house or enough money to send their children to school. The modality of migration therefore says little about decision-making or women's autonomy in migration.

This chapter has also shown that social networks are gendered and that women have lower access to what can be described as 'community-wide' social networks based on friendship and 'bridging social capital' (Pieterse 2003).

Women were also unable to access support from within the community in cases of domestic violence. Both the example of doña Paolina who escaped her violent husband from the mining town by going to Argentina in the late 1970s and the more recent example of Maria who was beaten by her brother for coming home late illustrate how women victims of domestic violence need to seek support from outside of their community. Social networks in these cases therefore 'fail' and show their gendered nature, given that at other times, such as don Serafino's daughter's death, the community is able to come together to support a man in crisis.

Because of the gendered nature of migration, the ways in which women and men use social networks also leads to a different geography of social networks between regional and global migrations. This chapter has shown that gendered migrations have contributed to this different geography of social networks. In Argentina, migrants migrated as a community and settled in three informal settlements in Buenos Aires. Migration to Spain was much more of an individual project, supported by the family. As a result, the geography of migration is much more dispersed. When organising fieldwork, I was given contacts from San Sebastián in the North of Spain to Tenerife in the Canary Islands. In the end, I settled for fieldwork in three cities, which were relatively representative of the types of places that Bolivians in Spain migrated to and where I had good contacts. Women were initially ostracised from the community social networks that people used to migrate to Argentina. However, they also built their own social networks and used them to migrate to Spain.

Notes

1 The term he used was *convivencia* which literally translated means 'living together'.
2 It is important to stress here that a conscious effort was made in order to understand these issues from the migrants' point of view, recognising their own agency. Therefore, while an 18-hour working day would be deemed abusive and exploitative by most people, I will here present it as such only if the migrants themselves perceived it and understood it as abusive.

5 Work

Work in the mining centre

In the mining centre, the 'ideal' gender relations were built around a strict notion of men being the breadwinners, the miners, and women the housewives. Housework and care work were seen as women's responsibility. Women raised children, cooked, and did the washing. They had to do all the washing by hand and often had to carry the water to their houses. Girls started helping at home from a very young age and were socialised to clean, wash, cook, and look after the children. Serving men – in the physical sense of serving food at the table as well as in the broader sense of making sure that they are fine – was women's responsibility. It was assumed that women did not engage in paid work. The paid work that they did carry out, such as petty trading, raising animals for sale, selling food, or knitting, was not considered 'work' as such but seen as 'supplementary income', as in other societies that are organised along patriarchal lines. Therefore, while the rigidity of this gender division of labour in relation to housework largely reflected reality, in relation to paid work it did not. Some women worked, not just in the socially acceptable role as *palliri,* or (usually) widowed women who work outside the mine, sorting valuable ore from the mining rubble, but leading *cuadrillas* or groups of miners, down the mine. Doña Claudia, who was born in 1955, recalls how she used to work in the mine from when she was 12 years old: "When I was almost 12 years old, I had my *ficha* [cooperative membership card], I used to work in the mine […] from when I was little, I used to work there". However, the work she did was different from the way men mined. Men used to use dynamite but women used to carry out the 'leftovers': "We used to take that out with a bag, or in a wheelbarrow. […] We didn't work with dumpers. In a month, we used to get 20kg, 30kg". They mined everywhere. If there was mineral in the mine, they worked there; if in the river, they collected the mineral in the river. Women and children worked in the river as well (Buenos Aires, 11th March 2003).

Sandra recounts how her mother used to work as a *palliri,* "those who break the stones," and that besides fulfilling her domestic responsibilities, she used to butcher pigs, prepare the meat, and sell it on: "in a year, she would butcher ten pigs, two or three times a year. She sold them on as sausages, *escabeche, fricasé, enrollado, chorizo,* different styles of pork." She remembers helping her mum, walking 20, 30, or 50 kilometres to buy the pigs from rural households and then walking back to the mining town with the pigs (Cochabamba, 25th May 2002).

Bolivia suffered from hyperinflation during the 1980s. The government, under advice from the International Monetary Fund, imposed a severe restructuring of the economy, known as the New Economic Policy 21060 (Kohl 2006). It dismantled the national mining company, Comibol, which led to the relocation of miners from mining centres to the cities. The price of tin had decreased during the same decade, which made mining unprofitable even for those miners working in cooperatives. So self-employed miners started moving towards the cities in search of better working conditions.

People from this particular mining town started moving to Cochabamba towards the end of the 1980s and early 1990s. This internal migration went hand in hand with an initial reorganisation of gender relations, as women had to find paid employment in the face of growing male unemployment.

Work in Cochabamba after internal migration

The move to Cochabamba entailed significant changes in the economic organisation of the now ex-mining households. First, men had lost their livelihoods. They found it hard to adjust to the urban markets, at least initially. Given the timing of their internal migration, in full economic restructuring, they found that there were already high levels of unemployment in Cochabamba. Women's role within the household was one of the first things that changed. Women started looking for work and entered the labour market in greater numbers. From a predominant 'male breadwinning' model, by 2008 the households in the neighbourhood had very similar labour market participation rates to other places in Bolivia: the sex ratio of the Economically Active Population (EAP) for the community was 40.2 per cent, which is only slightly below the national average of 40.6 per cent of all EAP[1].

Excluding students and retired people, almost half of all women (45.7 per cent) were housewives in 2008, a decrease of over 12 percentage points from 2002, when 57 per cent of all women were housewives. Of those who were economically active, almost half (45.7 per cent) worked in trade, followed by

Table 5.1 Labour market participation, by sex, 2008

Labour market participation, by sex, 2008, %			
Working status	Women (N=230)	Men (N=250)	Total
Employed	43.9	60.0	52.3
Not economically active	56.1	40.0	47.7

Source: Survey, 2008, cases above 16 years old

semi-skilled workers (10 per cent), unskilled workers and teachers (8.6 per cent in each, see Table 5.2).

The category 'other' for women includes: a driver, hairdresser, secretaries, garment workers or seamstresses, bakers, and an accountant.

This data excludes those who had migration experience or were absent at the time of the survey, to avoid having the results heavily skewed towards the occupations that migrants do abroad: care for women in Spain and garment work for both men and women in Argentina. However, this means that the figures presented here give a good indication of the structure of the labour market and the opportunities presented to this group of people in Cochabamba.

Almost a third (30.5 per cent) of all men worked in construction while a further 16.2 per cent worked as drivers. However, men had a wider range of occupations to choose from. Besides working in general unskilled occupations, they also worked in trade, as police officers, miners, teachers, and in smaller numbers as mechanics, farmers, locksmiths, artisan trades, bakers, electricians, tailors, accountants, shoemakers, painters, musicians, and engineers. Overall, men found work in 21 jobs while women concentrated in only 12 different types of job, which suggests they had less choice.

Migrants maintained and reproduced gender-based differences through the migration process. Women have fewer occupations to choose from in Bolivia as well as their countries of destination, as will be seen below.

From the mine to garment workshops: work in Argentina

As would be expected, the labour market insertion of Bolivians in Argentina is also gender-selective (Benencia and Karasik 1995; Orellana Halkyer n.d.). The literature suggests that Bolivian women in Argentina usually work in domestic work, trading, and garment manufacturing. Men, on the other hand, are usually employed in the construction sector and agriculture (Benencia and Karasik 1995; Benencia and Gazzotti 1995; Grimson 1999; Mugarza 1985;

Table 5.2 Current occupation: women

Current occupation: women, 2008	%
Trader	45.7
Semi-skilled worker	10.0
Unskilled worker	8.6
Teacher	8.6
Cook	5.7
Domestic worker	5.7
Other	15.7
Total (N=70)	100.0

Source: Survey, 2008

Table 5.3 Current occupation: men

Current occupation: men, 2008	%
Builder	30.5
Driver	16.2
Unskilled worker	10.5
Trader	4.8
Police officer	4.8
Miner	3.8
Teacher	3.8
Other	25.7
Total (N=105)	100.0[2]

Source: Survey, 2008

Recchini de Lattes 1988; Zunino 1997). This is similar to the labour market segregation by sex found in Asia where migrant women work in domestic work, entertainment, and nursing while men work in construction and transport (Abella 1995).

Taking both men and women together, 54 per cent of all of those who migrated to Argentina found work in the garment sector, 12 per cent in construction, and 10 per cent in trade. Only 4 per cent of those who migrated to Argentina worked in domestic and care work. On the other hand, of those who migrated to Spain, 52.8 per cent worked in domestic and care work and 33 per cent in construction (Survey 2008).

The differences become clear if we look at the distribution of jobs between women and men. For women migrants to Argentina, 50 per cent found work in the garment sector, 31 per cent in trade, and 12.5 per cent in domestic and care work, but in Spain, 94 per cent of all women worked in domestic and care work. For men, in Argentina, 56 per cent worked in the garment sector, 18 per cent in construction, but in Spain, 69 per cent worked in construction and then smaller percentages in other occupations.

Even in Argentina, women had a much more limited choice in relation to the number of occupations they could choose from. While men found work in 13 different occupations, women clustered in only five occupations. For women, therefore, the migration to Spain, while proving more lucrative, as will be shown, also implied, basically, no choice in the type of work they could carry out. There was just one option available for them: domestic and care work. On the other hand, men seemed to have more options available to them both in Spain and in Argentina.

A number of more upwardly mobile migrants were working in relatively more stable and better paid jobs, for example as bus drivers and nurses in Argentina or as engineers in Spain. Several also worked their way up the garment sector and became owners of garment workshops in Buenos Aires, as I discuss below.

Table 5.4 Occupation at destination, by sex

Occupation at destination, by sex, %

	Argentina		Spain	
	Woman (N=16)	Men (N=32)	Woman (N=18)	Men (N=13)
Domestic worker	12.5	0	94.4	0
Locksmith	6.3	6.3	0	0
Trader	31.3	0	0	0
Builder	0	18.8	5.6	69.2
Driver	0	3.1	0	7.7
Garment worker	50	56.3	0	0
Builder and garment worker	0	6.3	0	0
Baker	0	0	0	7.7
Waiter	0	3.1	0	0
Other	0	3.1	0	15.4
Security guard	0	3.1	0	0

Source: Survey, 2008

The garment sector[3]

Many Bolivians find work in the garment sector and have done so for decades (Bastia 2007; Benencia and Karasik 1995; Courtis 2000; Montero Bressán and Arcos 2016; Recchini de Lattes 1988). Bolivians have also started taking over some of the workshops, buying them from their Korean employers, and then using their networks to employ other Bolivians. Over the last three decades, we have therefore witnessed a third changeover in the garment sector that went from being predominantly Jewish-owned to Korean- and now to Bolivian-owned.

The garment sector was the most popular employment option for those who migrated from the *barrio* in Cochabamba to Argentina. Out of the total surveyed, 54 per cent had in fact worked in the garment sector, suggesting that it is a niche for the population of ex-miners (Bastia 2007).

There were some differences in how important the garment sector was as a source of employment for men and women in 2002. Half of all women worked in the garment sector as their main occupation while for men this was the case for only a third of them (a further 6.7 per cent of all men had worked in the garment sector as well as in construction). However, by 2008, the garment sector was employing half of all migrants, with a slightly higher percentage among the men, than among the women.

While working conditions vary from workshop to workshop, it is not uncommon to find appalling working conditions, which some label as being 'slavery like' (Montero Bressán 2011; Montero Bressán and Arcos 2016).

Migrants describe workshops as overcrowded, with limited lighting or ventilation. The fact that they are often not registered makes the working conditions worse, for example, by limiting the size of the windows or restricting the workers' movements. Many workers live at their workplace, sometimes taking turns to sleep on a bed shared with co-workers. Some parents are unable to leave their children in Bolivia and take them with them (see next chapter). This has led to tragedies, when these children become victims in workplace accidents, such as fires. In 2006, six people died in a fire in a garment workshop in Viale Street. Four of these were children ten years and under.

While the legal status of migrants improved dramatically after the passing of the new Migration legislation in 2003, the testimonies recorded during fieldwork in 2002 were full of stories of migrants being fearful of deportation. These deportations did not materialise. However, employers used them to limit their freedom of movement (see Bastia and McGrath 2011). Some workers did not leave the workshops for the first few months while some cases exist where they had no outside contact for over a year. Don Daniel, who travelled to Buenos Aires on various occasions, had Saturday afternoon and Sundays free, as is usually the case for live-in garment workers. However, he did not leave the house on Sunday: "On Sundays I stayed in. I didn't know the place. It's a big country. No, I couldn't go out. Besides, you had to have your documents [residency and work permit] and I didn't have them." (Cochabamba, 9th June 2002).

Given the tight work schedule and lack of alternatives, workers were also dependent on their employers for food. Employers offered two meals a day, but many complained that the food was inadequate. Isabel, a woman in her twenties who was trafficked to Buenos Aires at the age of 15, complained about the food she was receiving in the workshop where she was held. Describing an incident during which she and a friend of hers escaped, she said:

> She ran away first, then I ran away with my sister because she [the employer] was treating us badly, because the food wasn't as it should have been. The food, it was heavy, like here in Bolivia, noodles, an egg, something like that. That's why we ran away. [...] We had pasta, potatoes. [...] Meat is cheaper over there, but she [the employer] was lying to us, he was lying, telling us that things are cheaper over here [in Bolivia] than over there. But it was a lie. That's why we ran away. (Cochabamba, 18th June 2002)

Food is high on the migrants' list of priorities. Isabel was not the only one to give food and its availability high importance. Many other migrants value their ability to purchase better quality and diversity of food cheaper, relative to their wages, in Buenos Aires. Interviewees mentioned that being able to buy yoghurt, meat, or processed food, at the time considered luxury items in Bolivia, as one of the reasons contributing to the migrants' positive evaluation of Buenos Aires and their quality of life there.

Wages and their relative purchasing power were very important to migrants in their assessment of their experience abroad. The interviews indicate that men working in the garment sector were able to secure higher wages. Men were also more likely to be working piece-rate, following a short period of apprenticeship. For example, while the highest monthly wage for a man was 1,000 Argentine pesos (then equivalent to 1,000 US dollars), the highest equivalent for a woman was 500 pesos (500 US dollars). Both cases were working for Korean employers on piece rates.

Migrants also commented on the unequal wages paid to men and women working on a daily basis: 20 to 25 pesos per day for a woman but 35 to 60 per day for a man, both working for Korean employers. Migrants' experiences varied in terms of their migration status, the nationality of their employer and the actual tasks undertaken.

Working hours ranged from a 12- to a 17-hour working day. Working from 8am to 1am with a half hour break for lunch was quite common. The working week usually started on Monday and finished on Saturday midday, when most workers had to leave the workshop and therefore arrange their own food and housing. Wages varied greatly, depending on the experience, working mode, which included working by an hourly/daily/weekly/monthly pay or piece rate, and nationality of employer. While all returnees whom I interviewed had worked for Bolivian employers, those interviewed in Buenos Aires worked for either Bolivian or Korean employers. Only one worked for an employer of a different nationality: a Peruvian employer.

Among garment workers, there was a clear preference for Korean employers. They thought Koreans were more trustworthy. Bolivian employers often owed wages of several months. Some migrants were never paid, when their employers closed down their workshops, without paying their workers the wages they owed them. Many migrants started off working for Bolivian employers as a way of gaining the skills necessary for working in the garment sector and then progressed to working for Korean employers to gain better security and higher wages (Bastia and McGrath 2011).

Domestic work

Domestic work is in many countries the major employer of female labour, especially migrant female labour. It is seldom regulated and many of the workers' activities, especially when this involves migrant workers, are not categorised as 'normal' paid work (Anderson 2000). However, Argentina has recently passed a progressive law to regulate domestic work, Law 26,844. The new legislation recognises maternity rights and paid holidays and limits working hours to eight per day and 48 per week (ILO 2014). However, at the time the interviews were carried out, legislation stipulated that the working day for domestic workers lasted twelve hours and anything worked in excess of this could not be charged for. In case of illness, the worker could also be dismissed without having the right to ask for compensation, following a 30-day absence (Sociales and FLACSO 1995).

Domestic work as an activity presents many problems. The costs of working in this sector have been well-researched and include being on call 24 hours, low pay in relation to the hours worked, low status and no opportunities for improving one's occupational position (Anderson 2000; Ioe 1991; Chaney et al. 1989; Pappas-DeLuca 1999). In most countries, this is a feminised sector of the economy and usually involves a highly personalised working relationship. The usually very high differences in status between a domestic worker and her employer and the high incidences of sexual harassment have also been confirmed by the interviews (see below).

This special status of domestic work makes it highly undesirable as a potential source of employment. My interviewees therefore generally avoided working as domestic workers wherever possible. In Latin America, domestic work is usually undertaken by those members of groups with the lowest status within the national socio-economic and ethnic hierarchy. In most countries, domestic workers are women who have migrated from rural to urban areas and often of indigenous or African origins. It is not surprising therefore that in Argentina many domestic workers are migrant women. Paraguayans dominate this sector. However, it is also a significant source of employment for Bolivian women (Gogna 1993; Courtis and Pacecca 2010). Recent studies suggest that there is little social mobility in this sector and that, despite assumptions that this is a 'first step' job for newly arrived migrants, many women remain in this job for the duration of their working lives (Tizziani 2011).

For migrants however, and women migrants in particular, the domestic work sector presents certain advantages. Sometimes women are able to work in this sector while continuing to carry out the caring responsibilities they have towards family members. Live-in domestic work involves living in the employer's house. The employer offers a bed and food in addition to a monthly salary. On the other hand, in the 'live out' arrangement the domestic worker goes to the house to work for the day and returns to her own home for the night.

In a study conducted in Santiago in Chile, the live-in arrangement was found to be particularly useful for women who were migrating alone and for the first time (Pappas-DeLuca 1999). Despite the long working hours and being on call for 24 hours, 'living in' decreased the settling in costs usually associated with finding suitable accommodation for first time migrants. The worker could more easily save her monthly salary for other purposes because she had a bed for the night and her meals provided by the employer, as well as no bus fares to pay on a daily basis. However, this arrangement is limited to those who travel without their children and partners or husbands. It is therefore usually only open to younger women or to single mothers who were able to leave their children in their place of origin, usually with their parents or other female relatives.

Clearly, for many women the disadvantages outweighed the possible benefits. Monica, a woman in her thirties who was working as a geriatric nurse at
the time of her interview, travelled to Buenos Aires in 1998 and found work in
the domestic sector. She gave a vivid description of the way power relations
between her employer and herself made the job intolerable.

> My first job was domestic worker. [...] It was a total abuse, but well. [...]
> I tried working as a cleaner as well but it didn't work out. I couldn't. I
> went to a woman's house for whom I was a servant, a floor rag, not a
> person that was actually there. [I was] a machine that had to work. At
> one in the morning I was still cleaning and she never came close to me
> and offered a glass of water. We are as human as she is. Maybe she's got
> more money, but they treated me like that and I heard of other people
> who were treated in a similar way. (Buenos Aires, 23rd March 2003)

Monica had various jobs as a domestic worker and in most of them she
experienced some sexual harassment, by both men and women. With time,
she managed to secure a job in a geriatric hospital and train as a nurse. She
now has a stable and well-paid job as a nurse. At the time of her interview,
she was also studying for a degree in nursing with the aim of improving her
chances of being promoted and getting into teaching.

Construction industry

The construction sector is important for this particular group of migrants
given that building is, together with driving, the most common male occupation prior to migration. Almost a quarter of all economically active men
worked in the construction industry in Bolivia. However, only 18.8 per cent of
all male migrant returnees worked in the construction sector while in Buenos
Aires; this is mainly because my fieldwork coincided with the aftermath of the
2001 crisis, which hit the construction sector badly. I was unable to interview
migrants who were actively working in the construction sector at the time of
fieldwork, but I did find some among the returnees. This section is therefore
based on the information provided by those who had worked in construction
before the 2001 crisis.

Dario is a man in his early thirties who has lived in Buenos Aires since
1999. He was a successful builder in Cochabamba who started by working as
a *peon* or a builder apprentice and moved up the occupational ladder through
gaining experience and responsibility. Prior to migrating, he was earning over
1,000 US dollars a month, supervising various construction sites, and managing his own team of builders, but he ran into debt. When he arrived in
Buenos Aires, the most sensible thing seemed to be to try to find work in the
construction sector but he was unable to find work because he lacked a
document recording his employment history. Lacking these documents, he
was unable to find a job in the construction sector and therefore lost the

opportunity to capitalise upon the skills and experience he already had. Instead, he worked unpaid in an 'apprenticeship' for three months in a garment workshop owned by Bolivians before being able to find work in a Korean workshop where wages were acceptable to him.

Other migrants faced similar problems but were able to enter the construction sector, mainly because of the timing of their migration journey. Daniel migrated in 1994 and worked his way up the occupational ladder in the construction sector. However, he was forced to return to Bolivia after three years because of an increasing fear of crime and violence (Cochabamba, 9th June 2002).

The construction sector in Buenos Aires used to be one of the most important employers for Bolivian migrants. Respondents in Dandler and Medeiros' study (1988) reported that almost two thirds found employment in the construction sector, over half (56.7 per cent) employed as semi-skilled mason and day labourers. During subsequent trips many migrants progressed within this sector and by their third trip over half were employed as skilled masons and foremen, in addition to another 20 per cent employed in the lower-skilled category (Dandler and Medeiros 1988).

Working in the construction sector requires some adjustment, especially in relation to learning new words for the tools used. However, working conditions are deemed better than in the garment workshops, including better pay. Daniel, introduced above, worked in a garment workshop for the first three months, but when his employer refused to increase his wages, he found a friend who helped him get a job in the construction sector. His initial wage was 25 pesos a day, but this increased in commensuration with his responsibilities and by the third year, he was earning almost 1,000 pesos per month, which was then equivalent to US dollars (Cochabamba, 9th June 2002).

Antonio travelled to Argentina before the imposition of the one-to-one, when the Argentinian government pegged the Argentinian peso to the US dollar. He remembers construction contractors fighting over labourers. He recalls that then there were few Bolivians, but that employers thought well of them. The high demand for Bolivian labourers and the increase in rates of pay by contractors gave Bolivians the possibility to continually negotiate their weekly wage (Cochabamba, 18th June 2002).

Construction workers often slept at the construction site, to save money on accommodation and transport, and to allow them to work longer hours (Dandler and Medeiros 1988). Antonio also mentions changes in the working conditions. During his first trip to Buenos Aires, working hours were 9am until 5pm, but they changed during his subsequent trips to 7am to 6pm (Cochabamba, 18th June 2002).

Trading

Trading generally involves following *ferias* or mobile markets to different locations throughout Buenos Aires' marginal areas, including shanty towns.

Trading is largely undertaken illegally. Despite the fact that some markets have legal permission to trade, within the market itself there is usually a mixture of legal and illegal stalls. Migrants found it increasingly difficult to obtain legal residency documents. Lacking residency and working permits precludes them from trading legally and they therefore needed to resort to bribing the market organisers as well as police officers who supervise markets. This in itself involves some level of risk as the authorities have the right to enforce 'legality' through raids and they can confiscate traders' wares, should these lack the necessary permits. The geographical location where these markets take place and the lack of security means that the street vendors are at a high risk of being robbed. Trading therefore involves many risks and the life stories give evidence of the perpetual cycles of capital accumulation, followed by armed robberies, which effectively prevent migrants from planning, saving, and achieving their migration goals.

> Sandra initially moved to Argentina to work in a garment workshop owned by a Korean person. She worked from 8am to midnight for $350 a month. The working hours were too long so she later took up trading, buying, and selling clothes. Trading was quite a dangerous choice. She described how the *ferias* became very popular. People from the provinces started to come on Mondays and Tuesdays by bus. They arrived at 6 or 7am and the buses left around 10am. Therefore, those who were selling had to be there beforehand. They set out around 3am to occupy their posts by 4am. This also had its drawbacks. So much business attracted also robbers who would stop those who were going to sell there on the way to the market, at crossroads or at the red lights, point a gun at them and steal all their wares. If they refused, they would shoot them. As there weren't any police patrolling the area, they would go unpunished. Sandra got robbed three times, once with armed men. She lost $1,600 the first time and $600 the second time. The last time, she lost all her wares, when she left them outside locked up in a trolley. At this point, she gave up and returned to Bolivia. (Fieldwork notes, Buenos Aires, 14th March 2001 and 28th May 2002)

Trading serves an important role in women's labour market insertion as it gives them the ability to work in parallel with their responsibilities as main carers. In 2008, just over a third of all women included in the survey, who had been in Argentina, had worked in trading, but none of the men had worked in this sector. Women's life stories also show that they enter the trade sector when they are unable to work in either the garment or the domestic sector, due to their responsibilities towards young children. All interviewed women lost their jobs in the garment sector or domestic work after they became pregnant. Although some were able to continue working in the garment sector with their new-born babies, this became impossible when the babies started crawling.

Josefina undertook three trips to Buenos Aires, the last one in 1996 lasting four years. Her account tells of her successful migration to Buenos Aires. Capitalising on her previous trading experience, Josefina worked selling fresh fruits and vegetables in Liniers, where many Bolivians have opened their businesses, in the capital. She managed to set up her business, earned well, and saved. She overcame difficulties and managed to make good relationships with her neighbours. At first, she sold in the street on a stall but said that once she had her children she could not sell in the streets any longer and had to rent a shop. For part of her stay, she lived close to her shop in Liniers and then she moved to La Ferrere, another very 'Bolivian' neighbourhood in Greater Buenos Aires (Cochabamba, 8th June 2002).

However, not everyone is able to set up a successful trading business. Diana was also an experienced trader: she started to help her mum sell at the various local markets in Cochabamba at the age of six. When she travelled to Buenos Aires, she tried to get into the trading business but found it very difficult to make much progress. She did not like working for Bolivians because she felt they did not treat her well. However, she could not work for Argentines because she did not manage to get a work permit. Towards the end of her stay in Argentina, she became pregnant and had a baby, which made it even more difficult to find a job. She returned to Bolivia empty-handed and her mother-in-law gave her a hard time as a result (Cochabamba, 10th June 2002).

Women's relative success in trading and their labour market participation has repercussions for their relationship with their partners.[4] At one level, their participation in the labour market questions their partner's ability to support them. Their partners often use women's supposed primary responsibility for childcare as arguments against their participation in the labour market. Women's relative success at trading can also raise their partners' expectations, especially where women are solely responsible for their business. Elena's life story gives evidence to both accounts.

Elena was in her late twenties when I met her and she had a five-year-old son. She had been coming to Buenos Aires since she was 15 years old. Initially she worked in a garment workshop but later moved to a geriatric hospital, working double shifts for 300 pesos a month. She periodically returned to Bolivia and managed to finish high school. She then returned to Buenos Aires with the aim of saving enough money to enter university. She found a job in a garment workshop. However, she had already met her partner during one of her trips to Argentina. She felt that everything was going well for her until they started cohabiting and she became pregnant. She was still hoping to be able to go to university and was earning 400 to 500 pesos a month.

Her employer in the garment workshop let her work with her new-born baby, until he started crawling. Her employer then asked her to leave. Elena did not want to rely on her partner for financial support and tried to find work elsewhere, but having a small child and no access to childcare was an impediment for most jobs. Moreover, her partner did not like her 'overlooking' her mothering roles. Trade was her only option.

At the time, her partner was working for somebody else, making and selling *empanadas* (savoury pasties), so she encouraged him to start a business together. She wanted to buy an oven and the other things she needed to make *empanadas* and sell them. She even told him that she will pay him for the *empanadas* he will make for her to sell. Elena convinced him and she bought an oven. She worked for a month and was then able to afford a fridge-freezer, to make and store more quantities of the filling. Her husband prepared the ingredients and then left for work while she baked the *empanadas* and then went to sell them. She was successful, selling 100 empanadas on her first day.

Despite her best efforts to improve and expand their small business, her partner did not want her to work during the week, saying that demand was low during weekdays. However, he also opposed her looking for work elsewhere despite the fact that she was selling well. He insisted that she look after the baby. She found work in a grocery store, selling vegetables but did not tell her husband about it. She waited until he left the house and then left for her work. One day he returned from work early and found her working. Once discovered, Elena could not continue working in secret. She therefore had to obey her partner's wishes in terms of staying at home to look after her son. As soon as the opportunity arose, she started selling *empanadas* again and throughout her story it is clear that she was the one who insisted on expanding their production and coverage, thereby giving them both the opportunity to save and invest in new business opportunities.

Towards the end of 2001, Elena was offered a restaurant to rent. She saw this as the perfect investment for the money she managed to save through her *empanadas* business. They rented, furbished, and started serving hot food, pizzas, and drinks in a local restaurant. They were not making a big profit, but just breaking even for a couple of months until everything started going terribly wrong. They had borrowed some money to buy the furnishings and start their business. The economic crisis of December 2001 hit them hard, leaving them owing an increasing sum of money. They owed wages to their employers. They had also borrowed their capital in dollars so with the devaluation, this had tripled their repayment in pesos. Elena's determination to succeed in her trading skills backfired as her partner now blamed her for having lost it all: "Yes, I used to tell my husband: 'We are not the only ones', because he used to tell me 'Ah, you as my woman, you had to tell that we shouldn't have started that business.'" (Buenos Aires, 23rd March 2003)

Elena's account illustrates how women juggle around traditional understandings of gender relations and women's and men's respective responsibilities. Men sometimes use women's supposed primary role in mothering as an excuse for keeping them at home. Women often find ways to work around this, but then find themselves bearing the brunt of the responsibility should things go wrong. Women's engagement in trading creates tensions and raises expectations, but also gives women the opportunity to increase their bargaining power within the household.

These life stories illustrate that trading is a highly gendered activity in which women are able to capitalise upon their skills acquired before migration to maximise their profits abroad.

Trading in Cochabamba is traditionally a women's activity in which girls get initiated from a very young age (Dandler and Medeiros 1988; Balán 1995). If they manage to get their business established, this can give women a considerable degree of autonomy and success (Tapias 2015). However, this level of success is only achievable if they can ensure a minimum level of security and stability, which usually requires work and residence permits. Undocumented migrants were still able to trade, but remaining at the margins of the migration laws entails higher risks. This prevented undocumented migrant traders from saving and increasing their profits, which was the case with Sandra. Crime is also an ever-present problem. Many interviewees lost large amounts of capital due to armed robberies and almost all have witnessed this happening to somebody else.

The feminisation of regional migration allowed Bolivian women greater access to the Argentinian labour market and the acquisition of critical new skills related to migration: how to travel on their own, find accommodation, search for new jobs, tap into existing and create new social networks to help them achieve their objectives. Women clustered in a few occupations, but the work allowed them to challenge normative assumptions about gender relations. This also meant that they were increasingly responsible not just for the reproductive well-being of their families but also for their economic sustenance.

Spain

The new migration to Spain brought many new possibilities, particularly for women. Migration to Argentina, as we have seen, despite strong levels of feminisation, is generally perceived to have been led by men. There is, however, agreement that the new migration to Spain was led by women and that something changed profoundly in the structure of Bolivian society to allow women, including mothers, to migrate at such a scale (Interview with Bolivian sociologist, 15th May 2008). However, it could also be argued that the conditions that allowed such numbers of women to migrate had already been present for a long time. The context changed, in the form of new opportunities in Spain, and women were ready to respond. However, I would argue that the feminisation of regional migrations also paved the road for the strong participation of women in the new migration to Spain. In my sample, 57 per cent of those women whom I interviewed in Spain had previous experience of migration to Argentina. This is significant. Stories of 'pioneer' women indicate that they have all had previous experiences of migration within the region before embarking on their migration journeys to Europe.

Bolivians in Spain found a markedly different labour market to the one in Argentina but one in which migrant women faced significant labour discrimination (Parella 2003). In contrast with previous regional migrations, which were mainly led by men, the new destination, Spain, attracted larger numbers of women: about 57 per cent of Bolivians in Spain were women in 2008 (Escandell and Tapias 2009; Hinojosa 2008a). Many of these women found work in domestic or care work, usually looking after elderly people (Bastia 2011a). Domestic and care work is the first port of entry for migrant women, including Bolivians (Baby-Collin 2014; Gil Araujo and González-Fernández 2014). Spain has experienced an increased demand for elderly care work as a result of increased (native) women's labour market participation rates and very weak social protection policies for older people (Martínez Buján 2010).

Bolivian elderly care workers in Spain[5]

There were two main types of work arrangement for elderly carers: *interna,* or live-in, i.e. sleeping and eating at their employer's residence; and *externa,* working only part of the day and not sleeping at the cared-for house.

Most of the women interviewed worked as live-in elderly carers or *internas.* Live-in carers were usually on call 24 hours a day and had between a half and two days off per week. Their wages included boarding – accommodation and food – and varied from about 600 to 1,000 euros per month. None of my interviewees earned 1,000 euros, but they had been offered that wage or had heard of others earning as much. The advantages of working as a live-in carer were that they could send most of the wage home. Although live-in carers saved on food, accommodation, and travel, they usually still had to rent a room, which they often shared. This rented room provided a place to keep their belongings and spend their time off at weekends. Many interviewees took up additional jobs on their days off, such as a cleaning job that was paid by the hour. This meant that they had very little time to rest, sometimes only one night. However, they had to share sometimes very small spaces with the people they were looking after. One young migrant was not allowed out of the small $50m^2$ apartment she was living in with an 80-year-old woman, except once a day to empty the bins, despite also having a small baby (see Image 5.1, see also Chapter Six).

Externas generally worked only part of the day, usually 12-hour shifts and they also received a monthly wage. The advantage of having this work arrangement was that carers did not need to be on call 24 hours. They could also enjoy more time on their own and sometimes combined two jobs. Those working as *externas* usually earned 600 euros per month for 12-hour shifts six days a week, or just over two euros per hour. However, they often managed to have two jobs, a day and a night shift, earning up to 1,200 euros per month. This was extremely tiring and they were generally able to sustain this work arrangement for only a limited period of time as is illustrated by the interviews with returnees in the following sections.

Almost all interviewees worked informally and privately. There was only one exception, Cecilia, who worked for a private residence. Her better salary

Image 5.1 View from 50 m² flat where elderly carer worked, Spain

and working conditions reflect this. She worked night shifts in a private residence for terminally ill elderly people: two nights on and three off for 800–900 euros with all benefits included, such as overtime, holidays, and *media paga,* the two half wages given every six months in addition to the monthly wages. Undocumented status was an important shaper of working conditions. Cecilia, for example, saw an improvement in her working conditions after obtaining her residency and work permits, and after receiving her *media paga,* which she was not receiving before she regularised her status. Employers often do not pay out the *media paga* to undocumented migrant workers. Many interviewees identified their irregular status as the reason for their lower bargaining power vis-à-vis their employers, which meant that employers were able to get away with not fulfilling promises for pay or help with regularisation – commitments which they had made verbally.

I carried out the interviews during the financial crisis (May–August 2009), so it is unsurprising to find that all interviewees mentioned that conditions of work had worsened. Most were unable to negotiate time schedules. Women who held their jobs since before the crisis were generally able to maintain their wages, but those seeking new employment as cleaners were offered lower wages, sometimes as low as three euros per hour when the going rate before the recession was between eight and nine euros. This might suggest that carers had an incentive to stay with their current employer but as discussed in the next sections, monetary gain was not the only, or even the most important, factor for some elderly carers in their employment decisions.

Most women looking for new jobs mentioned having far lower bargaining power and potential employers being unwilling to negotiate working times, adopting a 'take it or leave it' attitude. This suggests that higher competition was making it difficult to find new jobs and negotiate their employment conditions. A representative of a migrant women's organisation interviewed in Madrid mentioned that a number of Bolivian women were offering to work for below the going rate. When faced with competition from other workers, some offered to work for 600 euros per month even when employers advertised jobs for 700 euros per month (interview with migrant women's organisation in Madrid, 7th August 2009).

Live-in elderly care work involves doing everything that is required (Twigg 2006) – as requested by the employer, the cared-for person, or as deemed necessary by the elderly care worker – to ensure the well-being and safety of the elderly person being looked after. Interviewees confirmed this in their descriptions of what their work entails: helping the elderly person get up, giving them a bath, putting on nappies, cooking, cleaning, making up the bed, mopping the floor, shopping and making sure that there was sufficient and adequate food in the house, taking the elderly person for a walk (see Image 5.2), feeding them, keeping them company, talking to them, keeping noise down when they are asleep, etc.

Image 5.2 Daily elderly care work, taking the 'granny' for a walk, Spain

Lucia, for example, when asked what her job involved, described: "Taking her out, put the washing machine on, I mean, housework, cooking, helping her with stuff she couldn't do, sometimes she would forget things" (Cochabamba 16th May 2008). While many of these activities could be understood as just tasks that elderly carers perform as part of their duties, many interviewees talked of feeling responsible for the well-being of the elderly people they looked after. They would put the person they looked after before their own well-being, for example, by never leaving them alone, or making sure that they found somebody trustworthy to look after them when they had to leave their jobs to return home. Emotional responsibility was therefore not linked to specific tasks but to the relationship that many elderly care workers established with the people they looked after.

Most migrants spoke in detail about their daily tasks involved in caring for elderly people, which challenges suggestions by Twigg (2006) that these are usually silenced and not talked about. For example, Fernanda, who was in her mid thirties, married, and had two children in Bolivia, said: "He was paralytic, I had to change his nappies, clean his bottom, I had to take him to the toilet, everything. I had to take him, change him, make him eat, that's what my job was like" (Cochabamba, 29th April 2008).

Some women worked as cleaners as opposed to carers. As cleaners they could work by the hour and combine work for multiple households. In this way, they sometimes managed to save a substantial amount in a relatively short time, if they managed to get enough work. Some also preferred to work as cleaners so they could combine child-care responsibilities and work when their children were in school or being looked after by their partners (see Chapter Six on care).

All the women interviewed in Spain worked as cleaners or carers, except for one who set herself up as a builder-cum-decorator, because she did not like carrying out the tasks involved in care work.

Construction

Men, on the other hand, worked in construction or (a minority) in agriculture. The construction sector was severely hit by the financial crisis. Interviewees talked of the insecurity associated with being undocumented, either on the basis of their own experience or those of their co-workers. Valentin, for example, who found work as a builder, despite not having any experience, explained that his employers preferred to hire Bolivians because they save a lot of money by hiring undocumented migrants. He said that "Spanish people always want to earn what is stipulated in the legislation" (21st June 2009). That is, they want to have their social security paid and work legally. Migrants, on the other hand, often accept working conditions that do not conform to labour legislation. He mentions undocumented co-workers being paid 30–36 euros per day on some construction sites.

One strategy used by employers during the aftermath of the crisis was to temporarily sack some of their workers or give them an extended unpaid holiday. Gregorio was claiming unemployment benefit at the time of the interview. He said that various people he knew were 'given holidays' during the summer of 2009 and not paid, so that companies would not go bankrupt because of the lack of work. The unemployment benefit he was able to claim was 1,000–1,100 euros per month, compared to his normal salary of 1,500–1,600 euros per month.

Men, such as Valentin, talked about feeling constrained by the lack of choice in jobs and the fact that his studies were not recognised in Spain (Madrid, 21st June 2009). Facing a labour market that only wants to hire workers because they are cheap contributed to this feeling of not having choices, not being able to use the skills he has, having to lie about past work experience to get the jobs that he did not want anyway. This contributed to stress and anxiety because he had to pretend he had work experience in construction, when he had studied something completely different: veterinary science.

This feeling of constraint in relation to jobs or their lack of choice is compounded by spatial restrictions that relates to their lack of documents. Valentin also talks about wanting to apply for citizenship "so I can come and go" (Madrid, 21st June 2009). He does not see Spanish citizenship as the final test of belonging to the Spanish nation or giving him greater opportunities to settle down and stay in Spain. Citizenship, for him, provides the means to live his life transnationally, to be able to come and go, and greater access to Bolivia and not having to renew his residency permits every two years.

Despite men's feeling that they did not have much choice, the evidence shows that women had even less choice. However, women tended to accept this limited choice (or no choice) in their attempt to make the most of the opportunities offered to them by the Spanish labour market, which, indeed, proved short-lived.

Does migration lead to social mobility?

Migration often leads to greater social mobility. Studies from the rural areas of Cochabamba identified migrant households as being significantly better off than their non-migrant counterparts (Yarnall and Price 2010). In a study conducted in the Valle Alto, remittances contributed 48 per cent of family income to migrant households, which was on average over 20 per cent higher than that of non-migrant households ($3,702 vs. $3,052 US dollars respectively) (Jones and de la Torre 2011). Similarly, evidence from the peri-urban neighbourhood suggests that migrant households (those that had somebody abroad at the time of the survey or whose members had previously migrated and had then returned by the time of the survey) are better off than non-migrant households.

Migrant households are more likely to be made out of bricks (71 per cent vs. 66 per cent); have a tiled roof (27 per cent vs. 10 per cent) as opposed to a corrugated iron or a flat cement roof (*losa*); and have the WC inside the house (30.2 per cent vs. 9 per cent). Migrant households also have an average total

Table 5.5 Money borrowed and saved, remittances, and income by destination, US
dollars

Money borrowed and saved, remittances, and income by destination, US dollars				
Migration destination	Argentina	Spain	Other Europe	Other
Amount borrowed	$717	$2,175	$1,500	$250
Migration income	$607	$1,472	$1,236	$962
Monthly remittances	$106	$436	$514	$85
Migration savings	$14,319.44	$28,545.40	n/a	$933.33

Source: Survey, 2008

income that is over twice that of non-migrant households ($645 vs. $304 US
dollars per month).

Migrants have a higher level of education: 10 years vs. 9 years for non-
migrants. This holds true for both men and women, although it is unclear
whether this difference is due to migration or is part of the self-selection
mechanism by which those individuals with higher education are more likely
to migrate. I have found a few instances in my qualitative interviews, where
migrants have been able to invest in higher education after a return from
abroad, but this was by no means everyone's experience.

There is also a hierarchy among migrants. Those who have been able to migrate
to Spain or other European countries have achieved higher levels of income,
remittances, and savings, as compared to those who have migrated to Argentina.
Monthly incomes in Spain were over twice the amount migrants earned in
Argentina ($1,472 vs. $607); as were the overall savings ($28,545 vs. $14,319).
However, the debts that migrants incurred to be able to migrate to Spain were three
times as high as those needed to cover migration to Argentina ($2,175 vs. $717).

In addition, among migrants, there is a more equal distribution of house
ownership between women and men, although men still outnumber women in
single owned houses by almost twice as many.

Housing is the most important objective for migrants. Most savings and
remittances go on housing (see Image 5.3 and Image 5.4).

Most migrants preferred to spend their savings on a house as opposed to investing in a business because houses could be built in stages, accommodating multiple
trips abroad, without the need to be present to nurture and supervise an incipient

Table 5.6 House deeds, by sex

House deeds on name of...	Migrant households (N=61)	Non-migrant households (N=60)
Man	49.2	70.0
Woman	27.9	20.0
Both man and woman	23.0	10.0

Source: Survey, 2008

Image 5.3 Three-storey house, Cochabamba

business. Many women also bought cars when they returned from Spain, for their husbands to work as taxi drivers (see Image 5.5).

Investment in taxis is significant, not so much from the economic point of view but more so for the symbolism that it entails. When I asked interviewees about the actual earnings made from the taxi business, most state that the income is very little. Driving a taxi entails very long hours or paying a driver to drive the taxi for you. Owners have to pay quotas to the taxi association and they often do not take into account the cost of car repairs and the rapid degeneration of cars driven on poorly paved or unpaved roads. The 'investment' into their husband's business became clear when one of my interviewees stated in passing that at least the taxi "gives him something to do". Others researching Bolivian migration have shown that men feel stigmatised when they are being supported financially by their wife who is working abroad (Parella 2011). While I did not come across such direct references to *hombres mantenidos,* or 'supported husbands', I understood the buying of the taxi as a symbolic gesture that women enacted to return the breadwinning role to their husbands.

Becoming breadwinners

Migration allowed women from the *barrio* to become the breadwinners of their families. They did this by inserting themselves in gendered labour markets and accepting precarious job conditions.

Image 5.4 House being built with remittances, Cochabamba

The gendered access to labour markets, combined with the gendered nature of labour markets, led to a segregated labour market insertion. Despite the precariousness of working without or with limited rights at work and often for low pay, as well as the ethnicisation of specific jobs, there were also some unexpected gendered outcomes. Women earned less than men in both Buenos Aires and in Spain, but in Spain they fared better during the recession. Despite worsening working conditions, many women were able to hold on to the jobs that they had, while men were much more likely to become unemployed. This is then reflected by the greater likelihood that men showed for returning to Bolivia and the higher percentage of women among Bolivians in Spain after the crisis struck.

Men and women also experience migrant work differently. While women often felt pride in what they could achieve with their savings and some found emotional connectedness in their work (see Chapter Seven), men found it more difficult to accept the low status of their jobs. In line with others who have researched such 'role reversals' of Bolivian women working abroad, I also found that women were proud of their achievement (Parella 2011; González and Sassone 2016). However, I cast doubt on the extent to which these changes can be classed as 'empowering'.

Compared to the mining town, there was a convergence of men's and women's working lives so that in many cases, both worked and in some cases, women took over the breadwinning role. These work trajectories enabled

Image 5.5 Investment in a taxi, Cochabamba

migrant households to accumulate significant benefits. Besides having higher incomes while abroad, migrants also often returned with large savings or remitted sums of money that are comparatively speaking much larger than the average local incomes. While house deeds continue to be unequally shared between men and women, with more houses in men's names only, there is some evidence to suggest that this is different in migrant households. Those households where there currently is or has been a migrant, are more likely to have shared house deeds or for the house deeds to be solely in the woman's name.

The inclusion of the 'return phase' of the migration cycle in this research has also allowed me to better understand what happens in couples when women return. As shown, migration leads to significant material benefits. However, while women might be proud of their achievements, they are also aware of pushing their changes to such an extent as to rupture their relationship with their husbands. Most downplay their achievements, by saying that "nothing much has changed". Women also undertake strategies to return their relationships to pre-migration gender roles, by investing in their husbands' businesses or by buying them a taxi to return them 'their' breadwinning role. There are additional changes that take place as a result of migration. In the following chapter, I therefore turn to explore what these changes mean for care and social reproduction.

Notes

1 The national data includes people 7 years of age and over, but here I include cases 16 years and over.
2 The percentages do not add to 100 due to rounding error.
3 Parts of this section were first published in Tanja Bastia, 2007, From mining to garment workshops: Bolivian migrants in Buenos Aires, *Journal of Ethnic and Migration Studies*, 33(4): 655–669.
4 There were no men traders from this community among the interviewees.
5 Parts of this section were first published in Tanja Bastia, 2015, Looking after granny: a transnational ethic of care and responsibility, *Geoforum*, 64: 121–129.

6 Care

Negotiating social reproduction in transnational social fields

The literature on care and social reproduction is heavily skewed towards the experiences of women migrants in South-North migrations, particularly those of domestic workers migrating from poorer countries in the Global South to richer countries of the Global North. Internal migration, South-South migration and the migration of men have been largely overlooked (Kofman and Raghuram 2015). It is impossible to address all these shortcomings in a single piece of research. Therefore, in this chapter I attempt to broaden our understanding of migration and social reproduction by also including South-South migration, the experiences of men as well as sectors of work other than care work. This approach muddies the waters a bit, given that we are no longer just focusing on one migration stream (South-North), one labour sector (domestic work) or one group of migrants (women). However, this initial scattering of ideas can eventually lead to richer and more complex analyses.

Care is a contested term (Kofman 2012; Kofman and Raghuram 2015). Some scholars argue that it is Eurocentric (Green and Lawson 2011); others, that it is not very useful because the same term is used for care practiced within the family as well as care carried out in a marketised setting such as a care home. These scholars argue that the setting within which care takes place matters and as such, research should pay more attention to these distinctions (Twigg 2006). Authors following this position highlight that family care is usually performed in a context of 'emotional connectedness' because it is given and received by family members. They argue that we need to distinguish this type of care from the care carried out in a work setting in which 'emotional connectedness' cannot be assumed (Twigg 2006). This presupposes that the best setting within which to practice care is within the family and does not always take account of the inequalities inherent in the fact that women take up the largest share of caring responsibilities. Both assumptions have been widely criticised, showing that care is a globalising concern (Robinson 1999) and that not all households are composed of heterosexual couples with children, or indeed, that familial relations are universally taken as being the most significant (Roseneil and Budgeon 2004). Moreover, families are not always spaces of harmonious and caring relations; while work settings can become spaces of emotional connectedness (Bastia 2015a).

Within a global setting, commentators have used care to analyse the inequalities embedded in the global circulation of domestic workers and carers. Following Arlie Hochschild, this type of migration can be understood as being part of a 'global care chain', or the giving and receiving of care at different points along the migration route (Hochschild 2000). Social scientists have drawn attention to these global care chains for their relationship to injustice and inequality, and the fact that most care work is currently distributed vertically down the social hierarchies of class and race (Hochschild 2000). They conceptualise care as a resource and argue that the migration of the main carer leads to 'care deficits' in places of origin. However, recent research on care from Latin America, has shown how closer attention to local care practices indicates that instead of a 'care deficiency', as is suggested by the care chains literature, the migration of the main carer results in a reorganisation of care that is not necessarily linked to a 'care deficit' (Herrera forthcoming, 2013; Yarris 2017). Moreover, the 'global' perspective can provide a useful starting point for analysing the re-organisation of care. However, the same approach also de-territorialises the analysis, given that it pays little attention to the specificity of the locations where some of these processes take place (Basch et al. 1993). A transnational perspective, on the other hand, helps focus on the specificities of these locations and how social relations, even when transnational, continue to be embedded in social hierarchies which are place-specific.

In this chapter, I build on the work on the care chains but focus more on exploring what migrants understand for care and how this influences their decisions to migrate, stay abroad, and return. Given that care is such a contested terrain in personal relationships, I also take the opportunity here to ask who cares, how they care, and who eventually benefits from this care. In answering this question, I first look at how care influences migration, for both men and women. I then discuss the challenges of combining migration with family life, including transnational parenting. In the second part of this chapter I look more closely at changes in housework as a result of migration and then touch upon migrants' parents, particularly grandmothers, who often take up caring responsibilities while their adult children are abroad.

Care is deeply embedded in interviewees' narratives of their journeys abroad. However, they have different conceptualisations of care and what it means to care for others. The form that care should take is often the basis of many migrants' rationales for undertaking their migration journeys, and those of their returns. Different understandings of what care means is also the basis for the tensions that exist between women's increasing wish to seek employment abroad and the resistance they face from others who see their plans as careless. This tension was most clearly seen in an interview with a couple in which Dario adopted a more traditional or one could say, conservative, understanding of care as requiring proximity and the involvement of mothers. Marina, on the other hand, argued that caring for her children involved her seeking work abroad, even when she did not have any contacts nor her husband's or her neighbours' support. In the face of open criticism from those closest to her, she nevertheless undertook the journey on her own, and left her children in the care of her husband (see also Chapter Four). For her husband, though, her choice was one of ambition and material improvement that went against his view that a mother needs to be close to her children. In a joint interview, he said so openly (see next section). She, on the other hand, argued that if men provided for their family appropriately, women then would not need to seek work abroad. Therefore, given men's inability to provide in a way that is acceptable to her, she undertook the journey to Argentina and was then planning another one to Spain (which did not materialise). She therefore also invoked a traditional ideal of family relations in her justification for seeking work abroad, but used it to point out that her husband failed to live up to this ideal (Buenos Aires, 7th June 2008).

In their interviews, there was discrepancy but very little ambiguity between their understandings of what it means to care, and as a consequence, also what it means to be a good parent (see next section).

Others, such as Javier, took a broader perspective and argued that families should not separate for any reason, no matter what their economic conditions and needs are, given that the consequences are irreparable:

> I think that families shouldn't separate for any reason, I mean, there is no valid reason fundamentally that would justify that a family separates, that the father leaves the children and the wife, that the wife leaves the husband and the children, I am convinced that there shouldn't be any economic justification or anything, because if that happens, well, the family will never be the same again. (Buenos Aires, 31st May 2008)

Instead of commenting on women's transgressions, Javier focuses on the family as a unit and places less emphasis on only women having to be present to ensure the well-being of the family members. Instead, his focus is on children and the couple being together.

This tension between men's and women's understanding of what it means to care and what can be included under that rubric can be said to be behind women's increased involvement in transnational migration for work. These personal and social meanings of care are just as important as the ways in which actual care is reorganised as a result of migration.

Care as a reason for migrating

Despite the fact that in the previous chapter I argued that work is the main reason why people migrate, it could equally be said that care is the fundamental reason for people's migration, except that migrants, particularly women, include work as being part of the 'caring' that a parent does. Most of the interviewees worked abroad and at a superficial glance, this is clearly a labour migration. When asked, most people said that they migrated for work because in Bolivia it was impossible to save, because they had debts, or they wanted to build a house. However, when one delves deeper into the migrants' stories, it becomes clear that having a house or some savings or wishing for a better future are all intrinsically related with caring. Migrants desire a house, or a better house, because their understanding of care includes the need to own a house, moving out of rented accommodation or improving the very basic houses of *adobe* and limited services, which is usually all they can afford with local unskilled jobs. Migrating to have some savings and wishing for a better future involve having some money aside so that one can keep the children in school and pay for their university education when the time comes. Labour migration is therefore often an act of care.

There were many instances where care work, whether paid or unpaid, was explicitly part of the main rationale for migrants' journeys. Maria, for example, was taken out of school and travelled to Buenos Aires when she was 17 years old, to look after her nephew while her sister worked in a garment workshop. She stayed for a year and a half. She described a very lonely experience of mostly staying at home with the baby, sometimes visiting a cousin or her sister-in-law, but avoiding going out because of the crime and costs involved. She also avoided leaving the informal settlement because she only had a tourist visa and was afraid of being caught by the police (Cochabamba, 18th June 2002). Others, such as doña Claudia, combined travelling for work and providing some childcare to her daughters who were in Spain. When her eldest daughter brought her own baby daughter to Argentina, doña Claudia helped with childcare, including picking her granddaughter up from nursery and taking her home so that she would not be on her own (Buenos Aires, 11th March 2003). At other times, she avoided migrating because she was looking after their grandchildren in Bolivia, while her daughters worked in Spain or in Argentina.

Men also migrated for care-related reasons. Some travelled to Buenos Aires to help family members with their business at times of ill-health. Others returned to Bolivia early because their elderly parents were ill. Vicente, for example, felt he had to return and help look after his two younger siblings when his mother was taken to hospital. When asked whether he would return to Buenos Aires he replied:

> Well, I wouldn't have left [Buenos Aires], but my mother was ill here [Cochabamba], she had been in hospital for two weeks and if anything was to happen to her, I have two younger siblings. Who is going to look after them? Who is going to guide them? (Cochabamba, 3rd June 2002)

When addressing care in the migrants' stories, it becomes clear that migrants manage competing understandings of care. Men in particular, but not exclusively, often adopt a traditional and narrow understanding of care, one that involves women (in particular) being physically close and present in the day-to-day activities of their children or other relatives, such as ageing parents, to ensure their well-being. This was the view of Dario when he explained, in a joint interview with his wife, that he had never wanted to come to Argentina and that he did because his children needed their mother close to them. Within the context of a discussion of a potential new migration to Spain, this husband and wife illustrate the competing understandings they have of what care involves:

> He: She knows. She knows very well. When we came, when she came from Bolivia, I didn't want her to; I had never wanted her to come here. I came for her, I had never thought of coming to Argentina. She came because of what they were saying, that in Argentina you earn so much, that this, that the other, for that reason she came. Well, life isn't like that. I came for the children, they were saying: 'Daddy, where is Mummy, go and bring her back'. They were little, my eldest was 5 years old, Dennis was four, and Javier was a baby. 'Mummy …' they were saying. Well, I had to come. I came and told her: 'Let's go back'. She knows. I never wanted to be here. 'Let's go.' And now she thinks … she always does this [she wanted to go to Italy]. I don't agree, how could I agree? She knows. My children are now older and they always need their mum close to them. Here it's either the money or the children. So I tell her: 'You decide'. I know her. I cannot stop her. Never. She knows … [silence] I know, but for me … [silence], also my eldest is in her youth, she needs a mother, Dennis is growing …
>
> Me: But many women leave, right? From the *barrio*, those who migrate the most are women, right, before it wasn't like this …
>
> She: Well, [before] they were like him [laughs]. I mean, that they didn't allow women to leave: 'that the children, our children, our daughter, that the children will be … they need …' and why can't a father give that, I say, right? Why, instead of women migrating, don't men put on their trousers and leave to seek what their children really need. I want something better for my children. That my children stay here in the *villa*, I don't want that. (Buenos Aires, 7th June 2008)

His wife clearly adopts a broader understanding of care, one that includes social mobility and 'something better' for their children. In their case, this includes moving out of the informal settlement. Others, like Sandra, also adopt this broader understanding of care. When asked about what her plans were, when she lived in the informal settlement in Buenos Aires, she replied:

> My plans? No, it was just to save and return to Bolivia, because, I mean, I was thinking about my son [her eldest]. I wasn't thinking about anything else, but my son. Clearly, it would have been nice to bring my son.

It would have been nice but no, take him away from my mum [who raised him as a baby], no. (Cochabamba, 25th May 2002)

As is already clear from some of the examples discussed so far, caring responsibilities are also often the reason behind migrants' return. Migrants often return to Bolivia because their children or other relatives fall sick and require care. The care needs are many and varied. Vicente returned to Cochabamba because his mother was ill and he felt he needed to look after his younger siblings. Alejandro, mentioned in earlier chapters, also returned from Buenos Aires and stayed when his father-in-law became ill during his visit back home, despite the fact that his plan had been to return to Argentina (Cochabamba, 9th June 2002). Marta also returned from Spain because her mother needed care and she felt that nobody but a family member would provide the right care for her. Sandra returned from Spain because her son needed an operation. Despite the fact that caring is definitely seen as a women's responsibility, these examples show that care concerns influence the migrations of both men and women.

Family separations

The family separations that Dario mentions are quite common, especially at the beginning of one's migration trajectory. While reconciling family and work life is difficult for everyone, migrants, particularly migrant women, face additional challenges given the structural position they occupy in foreign labour markets (Parella and Samper 2007).

Leaving your children behind was generally a prerequisite for travelling to Spain, at least initially. Among the 17 people interviewed in Spain who had children at the time of the interview, only three travelled to Spain with their children. Eleven left their children in Bolivia and of these, five managed to bring them to Spain and had children with them at the time of the interview. Of those who still had children in Bolivia at the time of the interview, five of the six were women and they all had teenage, as opposed to very small, children and four of these five women were single. Three interviewees gave birth in Spain. One of these sent her baby back to Bolivia to be looked after by her sister and mother, a practice that, according to the interviewee, was very common, although this is not supported by my data; one man initially had his children in Spain but then his children, and eventually his wife, returned to Bolivia. Therefore, for the migrants in Spain, it is much more common to travel there without their children and reunite their families later on, if possible.

Those with teenage children often opt for leaving them in Bolivia and single women in particular find it especially hard to reunite their families. This trend seems to contradict the general opinion in this community that it is easier to leave children behind when they are small because they do not understand as much and are more adaptable. Leaving teenagers, on the other hand, involves greater risks of them going astray (see discussion on when transnational parenting fails in this chapter). For the single mothers, it is imperative that they have good support from their families given the long separations. Even for those migrants who had migrated early on during the 2000s, regularising their stay took years and during this time, the relationship with their teenage children was maintained only over the phone or video-calls. Visits were not possible if they wished to maintain their access to the Spanish labour market.

Among interviewees in Buenos Aires, most had children with them. Among those interviewed in 2003, 17 of the 19 interviewees had children at the time of the interview, but only eleven had children at the time of their first migration. Of these, only one did not leave her daughter in Bolivia when she first migrated (Buenos Aires, 23rd March 2003). She did later on, when she went to Spain. There was very little change in 2008. Most had children with them (11 out of 15). Some of these had left their children in Bolivia at some point but then reunited their families (e.g. Marina). Some had taken their children with them to Argentina, but then left them in Bolivia with their mothers when they went to Spain (e.g. Monica). However, the overall pattern was for migrants to have their children with them. Only one man was working in Buenos Aires at the time of the interview while his wife and teenage children lived in Bolivia (Buenos Aires, 1st June 2008). This seems to suggest that once migrants are more settled and are able to consolidate their migration projects in Argentina, they are also able to enjoy family life and have their children with them.

Almost all migrants who were interviewed in Cochabamba had left their children in Bolivia for at least part of their migration trajectories, when they travelled to Argentina or to Spain. Among the 20 returnees interviewed in 2002 in Cochabamba, ten had had children at the time of their first migration and all, except two men, had left their children behind for at least part of their stay. Two women had taken children with them for only part of their stay. Three women also had had children in Buenos Aires. Among the men, three had left children in Bolivia, one had travelled with his wife and children, and one had travelled with a son and had already had two grown-up daughters in Buenos Aires. Among those returnees who were interviewed in Cochabamba in 2008, almost everyone had left their children in Bolivia. Among the 21 returnees, 12 had left all their children behind and one had left some children behind. Two had had children abroad and one of these had returned (from Argentina) when the baby was a few months old. One had returned (from Spain) when she was eight months pregnant. Three interviewees did not have children when they first migrated or when they returned (one of these had a baby after he had returned from Argentina because he had planned it that way). So among the returnees, only two had taken children with them but both were either unable to work or they found it very difficult to find accommodation as a result of having their children with them. The inability to look after their children while they were abroad is partly reflected in the fact that they have returned to Bolivia. These interviewees stressed the difficulties of keeping children, particularly in Spain.

The challenges of combining migration with parenting

While migrants do not explicitly state as a matter of course that they would have preferred to migrate with their partners, both men and women find it difficult to separate from their children. Ideally, they would want to migrate with their children, but this is often difficult or impossible. Migrants mentioned four main reasons for not being able to keep their children with them in their places of destination: high rents, work regimes, high maintenance costs, and in Spain, child protection legislation.

Rents

Despite the quite different contexts in which migrants find accommodation – informal settlements in Buenos Aires and flats in cities in Spain – the problems faced when trying to combine their housing arrangements with family life were strikingly similar in both countries.

When migrants first arrive to Buenos Aires, they tend to stay with friends or relatives (see Chapter Four) and after a short while, rent a room in a shared house in an informal settlement. Even those migrants, who initially stayed in other areas, eventually ended up renting in an informal settlement because of the closeness to places of work and the familiarity of being close to other Bolivians. Although everyone complained of violence and lack of safety in the informal settlements, only a few migrants actually moved out to a formal neighbourhood; while some in fact moved into informal settlements, having rented in formal parts of the city. Some could not afford to move to a formal neighbourhood and many chose to stay in the informal settlement. Free water and electricity and the familiarity of having Bolivian neighbours provided additional incentives for staying, although some mentioned wanting to move out of the 'villa' to be able to pay for services that they use, 'pay what's right', but this was an exception (Buenos Aires, 28th March 2003). At the time of fieldwork, there was also a lot of talk of formalising land titles in the informal settlement which provided additional incentives for 'holding on' and staying put (Bastia 2015b).

Renting in formal neighbourhoods in Buenos Aires is prohibitive to new migrants. To rent a flat in a formal neighbourhood, you require four months' rent in advance plus collateral in the form of a property owned by a close relative in the City of Buenos Aires (Bastia and Montero Bressán 2018). New migrants generally lack these prerequisites. They therefore either rent a room in a house in an informal settlement, usually owned by another migrant, or find a job in a garment workshop where employers provide food and accommodation.

As already discussed in Chapter Five, living in the garment workshop often leads to exploitative working conditions. Much of the literature on the labour exploitation of migrants, including trafficking, focuses on 'intermediaries' and recruiters. However, family members are often the ones who exploit migrants. This was the case for Irene, who went to Buenos Aires with all of her children to work in her uncle's workshop, but ended up earning hardly anything because she was charged for her children's food. They were also not allowed outside, as this was shortly after the Viale fire in which six people died and was followed by increased controls of garment workshops. During the interview, her sister intervened to say that they had to take their leftovers to their children to feed them and that their children were undernourished as a result. Looking for work elsewhere was difficult because they had the children and because of the extra controls following the Viale fire. Her husband therefore told Irene to come back to Bolivia and leave their children there (Cochabamba, 4th May 2008).

Migrants often resort to the extreme options of having their children with them in their places of work or leaving them at home with relatives – not all of whom will provide appropriate care for them – because of the difficulties of finding appropriate family accommodation. Many migrants complained about the exorbitant rents charged for accommodation for a room in informal settlements. Those who have resorted to living with their families – usually young families – in a room in a shared house complained of not being able to access the shared facilities when they needed them. The limited number of bathrooms needed to serve multiple families, often with young children, leads to long waiting times and inevitable accidents. Dario, for example, whose story of missing his children is told above, describes how his children had to use a bucket because their room was on the second floor, which was accessed through a winding stairway. There was only one bathroom for the whole house: "the children don't wait. We got them to wee in a bucket up there" (Buenos Aires, 2nd March 2003). It was too much for them, so they moved out. Despite the image of profit-seeking people who rent out rooms to fellow nationals, Dario stresses that the landlady was good. The housing conditions make living intolerable, not the people who rent out the rooms.

Rationed access to the kitchen means that some resort to cooking in their rooms, which then increases the chance of fires and other accidents. For all these reasons, migrants resort to splitting their families, at least initially.

Although the situation was quite different in Spain, the problems that migrants experience were very similar. Most migrants in Spain, as in Argentina, arrive at a friend's or relative's place. After a short while, they need to find their own space. This is usually a room in a shared flat. Despite the facilities being better, rents are often very expensive compared to the wages that migrants receive and many migrants resort to renting a room in a shared flat. Sandra and Fernanda, who travelled together, both shared a small room with Sandra's brother in a four- bedroomed flat shared between eight people. Their room was only big enough to fit two single beds and a wardrobe, with a space of about 0.5 metres between the two beds. When asked whether she thought about bringing their children with her, Sandra explained that it would have been impossible given the rent-to-wage ratio. She described the situation of a friend of hers who had two children with her in Spain:

> She had to earn 400 euros for the rent and another 300 or so for food, that's 700. Another 100 for eventualities. And the school, that makes 1,000 that she had to earn for rent, food, clothes, and children's snacks. So in all, you would need to earn 1,200–1,300 just to save 200 to 300 euros. And you have to slave it out and you need to educate your children, but you're not even going to be at home. You'll get home at night. Is that life, Tanja? That's not a life. (Cochabamba, 23rd May 2008)

Julia shared a room with a friend when she was single. She used to pay 280 euros rent for a shared room. Her salary was very low (600 euros per month) so she looked for another job at night. That way she was able to earn another 600 euros, making it a monthly salary of 1,200 euros, but she was working round the clock. This was not possible once her baby daughter was born, so she sent her baby to Bolivia:

> I sent her to Bolivia because I didn't have time to look after her. I want to work, I have to work. Here I don't want to have my daughter closed in a flat all day, or even worse, in a room. At the beginning, we were renting a flat and my daughter could crawl across the whole living room, but we were paying 700 euros, plus water and gas, making around 800 per month. But only my husband was working. It was enough to live on but not to save, but he didn't have his papers, and was walking around with the fear of losing his job, or worse, being deported, so we preferred to send my baby to Bolivia. (Madrid, 21st June 2009)

With one salary, it was impossible to pay all the expenses and save some money, which was the main reason for being in Spain. She had considered working, but could not find suitable childcare. To be able to access formal childcare provided by the council, she had to have legal residency and her employers had failed to provide her with a contract that she needed to start the paperwork. She mentioned informal childcare, but this is unregulated and unsafe, and the people who run it do not have any formal qualifications: "If I work I have to take her somewhere so she is looked after, but she's going to be closed in and I'm not going to know how she's being looked after. Those *miradoras* [childminders] aren't legal. [...] I don't have papers so they won't give me a place in a nursery. [...] We usually take our children to flats, rooms, but they are closed in, it's not monitored so it's a big risk" (Madrid, 21st June 2009).

She highlights the issue of space, comparing the space that her baby has access to in Bolivia and the enclosed living in Spain: "Now we're living in a room again so we can't have her with us. [...] In Bolivia, she lives close to a park so she has space to run around, but not here" (Madrid, 21st June 2009).

Long working hours

The long working hours that most migrants have to endure also make it extremely difficult to combine migration projects with parenting. Sandra has already hinted this in the quote cited above. Her own experience confirms this. While she was in Spain, her sister-in-law had decided to bring her son to Spain but finding it difficult to look after him, he ended up living with Sandra in Algeciras. He was already used to living with her because she had looked after him in Bolivia. When he lived with her in Spain, he used to go to school at 8:30am until 4pm, then was home on his own until 8pm or 9pm and when she arrived he was almost sleeping. On Saturdays he was on his own all day. Although this was doable, she was not able to look after him properly, provide guidance and proper care. They therefore decided to send him back to Bolivia, where he lived with his father, grandmother, and uncles. She also mentions this in relation to her own children, that she would not have had time to educate them: "How was I going to educate the children there [in Spain]? They would have been really *malcriados* [unruly]. I would have had to work three times as much. Moreover, children there enter in the materialistic stage and they want everything. They want everything. So, no" (Cochabamba, 23rd May 2008).

Although in Buenos Aires interviewees worked in different sectors, mostly in garment workshops, the conditions and long working hours were very similar, if not worse, as discussed in the previous chapter. Despite being geographically closer and the travel between the two neighbouring countries being much cheaper, the working conditions often enforced family separations for migrants, at least initially.

Maintenance costs

Rents are clearly the highest cost that migrants face when in Argentina or in Spain. While maintenance costs are higher in Spain for everyone, the fact that migrants mention these in relation to children refers to the fact that they are weighing up the alternative of having their children in Bolivia, where maintenance costs such as food and clothes are inevitably much cheaper. These costs are therefore not absolute but always relational.

Child protection legislation

Migrants are also concerned about child protection legislation and are aware that they are not supposed to leave children on their own. This was a specific concern in Spain but was also mentioned at times in Argentina, not so much in relation to leaving children on their own but by mothers who took their children to work and suffered accidents. Monica, for example, used to work in a restaurant and used to take her daughter with her. One day her daughter fell on a grill and suffered some burns. Customers advised her to stop taking her to work because social services, if informed, would take her away from her (Buenos Aires, 23rd March 2003).

Transnational parenting

Those who were unable to combine their migration projects with having their children close by resorted to what Hondagneu-Sotelo and Avila term 'transnational mothering' (Hondagneu-Sotelo and Avila 1997), or what I would term 'transnational parenting', to take into account the fathers who also parent from afar (Pribilsky 2007).

'Granny fostering' is a well-established practice and refers to women's leaving their children in the care of their mothers or other female members of their families for the duration of their stay abroad or in the city. This is common in most Latin American countries in both Central and South America and is not a new practice linked to transnational migration. It was quite common for internal migrants to leave their children in rural areas when seeking work in cities. The fact that they do not have dependants with them in the host country means that they can access employment opportunities usually available only for young single women, such as live-in domestic service (Pappas-DeLuca 1999).

Some have argued that Bolivian migration to Argentina is essentially a family migration. Cerrutti found that only 7.6 per cent of Bolivian women in Buenos Aires who have migrated a maximum of seven years before the survey was carried out had left at least one child in Bolivia (Cerrutti 2009). However, others have found that among their interviewees, 60 per cent had children, but of these, only 30 per cent had brought their children to live with them in Argentina (Correa and Pacecca 1999). I also found that most interviewees had left their children with family members in Bolivia, at least initially.

'Granny fostering' is widely practiced among women from this community (Bastia 2009). Most children were left in the care of relatives, usually the maternal grandmother, but there were many instances where children were left with aunts or the paternal grandmother. On some occasions when mothers migrated, their husbands were not capable or unwilling to take on the responsibility of looking after their children, so they employed the services of a neighbour or another person to look after their children. This was the case for doña Claudia, who started travelling to Argentina when her children were still little and paid a neighbour to cook and wash their clothes "like a grandmother [...] because children cannot get home to an empty house, because then they wouldn't study. So she was always there" (Buenos Aires, 11th March 2003).

Despite the fact that migrants felt sadness at leaving their children, the actual practice of granny fostering, where children are left with one's parents, was also practiced in the mining town. Sandra, for example, left her eldest son with her mother when he was little. She initially brought him to Cochabamba with her, but she realised that she could not work with the baby so her mother took him back to the mining town (Cochabamba, 25th May 2002).

Transnational parenting took multiple forms: from sending money and clothes regularly to the person who was looking after their children, to regular communication through phone calls and Skype, to regular visits by those who were able to secure their residency and work permits (see Image 6.1).

Catalina used to call almost daily, despite the fact that she knew that her children were well looked after by her mother (San Fernando, 11th May 2009). Roxana also talked regularly with her daughter, who was 13 or 14 when she left her in 2006, three years before the interview. She found that at the beginning, she was OK about being far from her daughter but after a few months, it became more difficult. However, she said that communication improved just before the interview and they were in touch more often. She felt close to her daughter and they were able to confide in each other. "… now we talk more regularly, my heart is fuller, I tell her 'I feel closer to you'. We always call each other. […] She tells me 'you are my mother, my friend, my *compañera*' so I talk to her but I always get very sentimental when we talk over the internet (with video-calls)" (Algeciras, 2nd August 2009). Therefore, they resort to phone calls as opposed to video-calls.

Image 6.1 Locutorio, internet café, accessing digital technologies to stay in touch, Spain

Visits

A key difference between those in Argentina and those in Spain clearly relates to documentation. Migrants in Argentina are able to come and go as they please, particularly after the 2004 migration reform. Some of those who had migrated to Argentina during the 1990s had difficulties similar to those faced by Bolivians in Spain, when planning visits back to Bolivia. For example, Marina had to travel with her youngest daughter's document, who was born in Argentina, because it gave her the right to be there. Her husband Dario could not travel to visit his children because he did not have a valid document. Sandra's aunt also mentioned that she brought her children to Buenos Aires only after she regularised her stay.

However, even before 2004, migration to Argentina was cheaper and easier. Some migrants had to resort to crossing *por río* or across unauthorised borders (see Chapter Four). However, migrants in Spain did not have these options. Visits home to either take their children to Bolivia or visit them if they were living in Bolivia were impossible for migrants who did not regularise their stays. Separations for those in Spain were therefore much longer. Four years were not uncommon.

A key objective for migrants was to regularise their stay so that they could come and go as they pleased, which often included visiting their children more regularly. One common remark for migrants was to state that they wanted to regularise their stay, so they could go back to Bolivia. Being in an irregular situation meant that they were trapped in Spain, as leaving would mean not being able to come back. Andrea, for example, left her daughters when they were teenagers and it took her four years to see them again: "Yes, I saw them again after four years. Last year I stayed three months, because I managed to get my papers so I stayed longer". At the time of the interview she was planning to stay five months because her youngest was expecting a baby. She feels bittersweet about these changes: "I won't always be able to be with my daughters, they already have their life. What can I do about it? I wouldn't have wanted it this way but ..." (Algeciras, 3rd August 2009). She was also planning to apply for Spanish citizenship.

When transnational parenting fails

Transnational parenting strategies were not always effective. Parents commonly complained about their children's not being fed properly. Sandra left her children with her mother in Bolivia while she worked in Spain. When she came back two years later, she found her children undernourished: "My reality, that's what was waiting for me. The children were undernourished. But once I got here, I gave them milk, food, now they're fine" (Cochabamba, 23rd May 2008).

Marina and Dario, mentioned above, left their youngest daughter in the care of the maternal grandmother when she was four months old. In his interview, Dario mentioned how painful it was to hear from others that they had seen their youngest daughter begging for milk.

> I was never thinking of staying. I brought them so they would stop suffering. I saw, they were telling me how my children were suffering in Bolivia. I was sending [money] to my mum. They were with my mum and my sisters. Maybe my sisters weren't buying them stuff but I was still sending them money. That's why I was working. So some people were telling me 'Your children are walking barefoot, they are begging'. I didn't like that. I preferred to bring them here. I'm more at peace. Here things are cheaper. At least here they also give you milk for each child. (Buenos Aires, 7th February 2003)

Marina travelled to Bolivia while Dario continued working. When she arrived there she confirmed that the story was true. So they decided to bring all their children to Buenos Aires. He had always been against their migration to Argentina and in his latest interview was quite explicit that the mother should always be at the side of their children.

Migrants felt that it was easier to leave younger children, because they were less aware of what was going on. Soraya, for example, said that it is easier to leave children when they are small because they do not understand, but teenagers have their own opinions. Her own children were teenagers when she left to work in Brazil, and they wanted to have her close by (Cochabamba, 25th April 2008). This was the same with Fernanda, who had a teenage daughter and a younger son when she travelled to Spain:

> My daughter, since I had been away for only a short time, she was fine, but was going out a lot. When a mother isn't at home, the children always do as they please. They lie to their grandmother. They say they are in one place but they go to another. My son was little so he was fine. I was more worried about my daughter. (Cochabamba, 29th April 2008)

There is a lot of talk of migration leading to teenage pregnancies as a result of lack of parental supervision. People tend to blame this on absent mothers and indicate that this is a new phenomenon, specifically linked to migration. However, teenage pregnancies were common even before members of this community started migrating internationally. This does not mean that they do not happen, though. In fact, a number of interviewees who had left teenage children when they went to Spain or to Argentina found out that their teenage daughters had become pregnant during their absence. Sometimes this led to them deciding to return to Bolivia. These migration circuits are closely interwoven with new ones. Doña Fabiana, for example, came back from Argentina because her daughter, then aged 17, got pregnant. Her daughter gave birth, finished high school, and then went to Spain, leaving her baby with doña Fabiana. At the time of the interview, doña Fabiana's daughter had another baby in Spain but was able to work and keep the baby, at least for the time being (Cochabamba, 5th May 2008).

Similarly, Andrea, who was introduced above, left her teenage daughters with her brothers and sisters when they were 17 and 15 years old. She had separated from her partner back in 1997 and he was not involved with their children. Andrea talked of the difficulties of watching changes from afar:

> There is little control from the father or the mother. If it's not your child, there is only so much that you can do [in terms of disciplining them]. I was lucky because my sister was strict with my daughters but even then, the youngest got married at age 21 and the oldest is 23 and she already has a baby. Maybe if I had been there, maybe the same would have happened. There are worse things in other families, so they are a bit left to themselves, when they are alone, there is nobody who tells them 'do this, don't do that', and they start doing certain things. (Algeciras, 3rd August 2009)

Don Tomas, interviewed in Buenos Aires in 2003, initially left his children in Bolivia and eventually brought them to Buenos Aires, save but the eldest who wanted to stay in Bolivia. His daughter was behind in school and was only just learning to read at the time of the interview, which he attributed to the fact that he left her in the care of his mother, who does not read or write:

> She is a bit behind with her speech and she doesn't read or write, nothing. She only learnt recently. She is starting to read now, perfecting. She used to say 'I can't, I can't!' Of course, she couldn't and nobody taught her, you see, we parents were working here [in Buenos Aires]; my daughter was there with my mother who doesn't read. So she couldn't [learn]. (Buenos Aires, 21st February 2003)

However, many interviewees were equally concerned about the upbringing that their children would receive, should they stay, particularly in Argentina. Most interviewees, with just two exceptions, rated the Argentinian educational system quite poorly. In their experience, Bolivian children were always *abanderados* or they came as top in their class, when they first moved from Bolivia to Argentina. They also complained of discrimination in Argentinian schools, particularly for those children who were darker skinned and more easily identified as Bolivian. For others, 'losing their way' was a reality despite having brought their children with them in Buenos Aires:

> We plan to go back. Here I don't like the way my children behave. In Bolivia, there is more respect. Here children lose their way. Twelve, thirteen they are already 'awake', but in Bolivia it's not like this. I don't know, I am surprised, when I was 15, 18, I didn't know what a boyfriend

was, or going out to dance. Now at 13, 12, they are already going out. That's why I don't like this country. (Buenos Aires, 6th February 2003)

Changes in gender roles during separations

As discussed in the previous chapter, women migrants at times took on the role of main breadwinner for their families. This implied separations over significant lengths of time, sometimes months but more often years. How did their families adjust to this new set-up in terms of daily household chores? The sexual division of labour within the home is a good indicator to judge whether there has been a reconfiguration of gender roles. As is well known from studies elsewhere, women continue to be responsible for a larger share of domestic tasks, even in cases where they have paid jobs outside of their homes (Hochschild and Machung 1989). In higher income countries, for which data exists, there has been a convergence between the time that men and women spend on typically feminine housework tasks such as cooking and cleaning. However, men's uptake of housework has been less steep than women's decrease in the hours they dedicate to housework. Moreover, there continued to be a gender gap in housework at the end of the first decade of the 21st century (Altintas and Sullivan 2016). Life stage contributes to some variability in who does what within the home. However, overall, even in couples where both partners work, men do fewer hours of housework across all life stages (Horne et al. 2018). Others, drawing on longitudinal, panel data-sets on time use in the US, have concluded that by "2009/10, women are estimated to do 1.6 times the amount of housework as men, on average (with wives averaging 1.7 times the housework of husbands, and married mothers averaging 1.9 times the housework of married fathers)" (Bianchi et al. 2012: 56).

In the *barrio,* as we have seen, women were socialised to be the carers and the home-makers. However, the survey carried out in 2008 shows that there is a difference in who does housework in those households that have engaged in migration, when compared to those who have no migration experience. Table 6.1 shows that in households where somebody has engaged in migration, there is a shift away from women and children doing specific household chores, towards fathers, other female relatives, and paid domestic workers taking up some of the tasks. Given the methodology and the small sample of households, it is impossible to generalise further. However, the findings indicate that there is a potential shift towards a more equal distribution of housework tasks resulting from migration.

Combining migration with family life

Travelling as a family is quite unusual in this group of migrants. As indicated above, even those who had their families with them during the interview, usually started off with one parent migrating first, then being joined by another and then by the children. Travelling all together implies high costs

Table 6.1 Sexual division of labour, households by migration status

Sexual division of labour, households by migration status, %

	Migrant households				Non-migrant households			
Person mainly responsible for:	cooking lunch N=67	shopping N=67	sweeping N=65	washing clothes N=63	cooking lunch N=77	shopping N=76	sweeping N=76	washing clothes N=75
Mother	71.6	65.7	43.9	44.8	80.5	59.2	50.0	52.0
Father	11.9	20.9	10.6	9.0	5.2	31.6	1.3	2.7
Daughter	7.5	3.0	18.2	3.0	9.1	3.9	22.4	8.0
Son	1.5	0.0	15.2	0.0	3.9	0.0	19.7	0.0
Other female relative	1.5	1.5	3.0	1.5	0.0	0.0	1.3	0.0
Paid domestic worker	3.0	1.5	1.5	1.5	0.0	0.0	0.0	0.0
Everyone	1.5	3.0	6.1	32.8	0.0	2.6	5.2	36.0
Mother and father	1.5	4.5	1.5	1.5	1.3	2.6	0.0	1.3
Washing machine				6.0				0.0

Source: Survey, 2008

and particularly for going to Spain, high risks given that visas were not granted in advance and there was always the possibility of being deported upon arrival to Spain.

Sonia is one of those who insisted on all going together. Her brother, who was already in Spain when she was in Bolivia, kept telling her to go to Spain on her own but she could not leave her children behind. Therefore, she travelled with her husband, their ten-year-old son, and an eight-month-old baby. They had always taken it in turns to work. She worked in Argentina, while he looked after their eldest son and also a nephew. He then worked in Bolivia, while she stayed at home. However, in Spain, they both had to work. Therefore, their ten-year-old son helped look after his eight-month-old brother (San Fernando, 29th July 2009). However, she was the only one who travelled to Spain with her family. The other two families who were reunited in Spain and did not leave their children behind with other relatives, adopted the strategy of one of the parents travelling to Spain first and the other one following a year later.

Horacio's wife went to Spain in 2004 leaving all their children behind. Some were in their twenties, but they also had a small one who was looked after mostly by their third-eldest daughter. In 2006, she came back and took the little one to Spain so he said, he "had to follow". Now they take it in turns to work. She works days and he works nights so they both look after their one-year-old (San Fernando, 27th July 2009).

Most migrants cannot afford childcare, not even informal childcare, as already discussed above. Those migrants who have their families with them resort to working shifts in turn to be able to look after their youngest children. Alternatively, one of the parents works only limited hours, while children are in school.

Rosa's husband went to Spain first and she followed a year later, with all their children. She is able to combine work with parenting because she is able to take her youngest with her to work, where she does nights looking after an elderly woman (San Fernando, 27th July 2009).

For some, having children abroad proved to be a blessing. Angela, for example, a young migrant who worked as a live-in carer, had a baby a few months before the interview took place and despite her change in circumstances, she was able to continue working. Her living conditions were in stark contrast with the happiness she felt for having her daughter. She lived in a flat, 50m^2, with a bed-ridden 80-year-old, whom she cared for 24 hours a day. Until two days before the interview, she was not allowed to leave the flat, except once a day to empty the rubbish bin. Under doctor's orders, she was then allowed to have a two-hour stroll with her daughter to get some fresh air and meet up with friends and her partner. When I asked her about her daughter, she replied: "I am so lucky. On the 3rd of January, they let me continue working with my daughter." However, having a daughter also implies having to stay in Spain for longer … "Now I have to build a house for us, for my daughter, myself, and my husband" (Algeciras, 4th August 2009).

Valentin and his partner also had a baby a few months before the interview. They had known each other in school but only started going out together when they met again in Spain. When asked if something had changed since they had their daughter, he replied: "yes, she gives us so much happiness; before we used to argue a lot but now not any more [...] she changed us" (Madrid, 21st June 2009).

Combining care and work is to some extent easier in Argentina because the migrant community is more established, so those interviewed in Argentina had lived there for a longer period of time. They had therefore been able to overcome the initial difficulties of combining paid work and looking after small children. This was also easier because of the geography of this migration, which meant that in Buenos Aires, families and friends travelled to the same neighbourhood, while in Spain they are more scattered across the whole country (see Chapter Four on mobility). However, even in Argentina, many interviewees talked about not being able to work, at least for some periods, while their children were small.

Three interviewees from 2002 interviews in Cochabamba got pregnant and gave birth in Buenos Aires (Sandra, Josefina, and Diana) but then went back to Bolivia. Both Sandra and Josefina managed to work, despite having babies – Sandra was selling at street markets and Josefina shifted from selling vegetables at a stall to renting a shop so she could keep her daughter with her. Diana on the other hand had to go back because she could not continue working in the garment workshop. She went back to Bolivia when her daughter was three months old (Cochabamba, 10th June 2002). She separated from her husband a couple of weeks before the interview. Marina, when interviewed in Buenos Aires in 2003, describes how hard it was to work with her baby daughter.

> I was working with her [until she was four months old, when she took her to Bolivia]. I left my room at four in the morning until nine at night. All day. Sometimes she cried, whenever I had to attend to customers, she always cried. They used to tell me off: 'why don't you leave her at home, just so you save the money [of paying somebody to look after her]. I replied that if I had to pay somebody to look after her, I wouldn't have much left. 'Sundays you have to leave her because she won't let you work'. So Sundays I had to leave her. I left her with a lady and she looked after her. I used to pay her. I was earning 15 pesos, I paid her five so I had ten pesos left. (Buenos Aires, 7th February 2003)

Others decide to go back to Bolivia to give birth. This was the case with Alfredo's partner, who recently went back to Bolivia from Spain to have their first baby (Cochabamba, 20th June 2008). In addition, Lucia, who was in high school when she was sent to Spain by her mother to work to help repay a debt and who returned to Bolivia three years later expecting a baby; as with Angela and others, for Lucia, her pregnancy and starting a family of her own also involved planning future migration journeys to help secure her son's future. "I was thinking of going back to Spain. I was thinking, my mum also told me, for his future because the money I brought from there isn't enough because she will need more" (Cochabamba, 16th May 2008).

Migrants or their partners travel back to Bolivia to give birth because they are uneasy about having their children in Spain. Some of the testimonies related so far already give an indication of why this might be the case. Julia's experience described above shows how difficult it can be for migrants to combine parenting and work, especially when they are also unable to afford renting a sufficiently large space to allow children some room to play and move around. Antonia, who travelled to Buenos Aires the day after she got married, and was eight months pregnant, is another example. She was 'being called' by her uncle who needed help with his grocery store. Being underage, she decided to get married to be able to travel because she thought she would not be able to cross into Argentina. She started working with her husband at her uncle's grocery store a few days after she arrived in Buenos Aires. After having her baby, she took some rest but then started working again, under increasingly strict control from her uncle and aunt, to the extent that she had to stop breastfeeding her baby against her wishes, because her uncle and her partner thought that breastfeeding was taking too much of her time and was keeping her from working efficiently. Antonia and her partner ended up having to escape from her uncle's grasp to come back to Bolivia. Instead of saving some money, they ended up owing her uncle money for food and lodging given that he was noting down any expenses, including those he made without their explicit consent (Cochabamba, 9th May 2008).

The migrants' parents

The children are not the only ones who might be 'left behind'. The migrants' parents are often left out of any discussion about social reproduction given the overwhelming concern for how children fare when their parents, particularly their mothers, migrate. In most Latin American countries, women traditionally provide not only childcare but also elderly care. When they migrate, they often continue to care from abroad, by sending remittances, making telephone calls, or paying for hospital bills. This creates complex circuits of care, yet in most of the literature, the elderly have often been rendered invisible, unless they are present as providers of care for their grandchildren (Vullnetari and King 2008; Bailey 2013).

The migrants' parents are also sometimes in a vulnerable situation. The severity of the vulnerability will depend on the migrants' socio-economic status, their parents' age, their health, and on whether they have additional caring responsibilities which have resulted from their children's migrations. As seen above, when migrants leave their children behind, they generally leave them with their own parents – the children's grandparents – preferably with the maternal grandmother. In this community, men often die prematurely due to their past work in the mine. Those who worked in mining often die in their thirties or forties, so the elder population is skewed towards women (Bastia 2009). However, the grandmother's ability to look after her grandchildren will depend on her age and fitness, health, and her financial standing.

I interviewed a small number of relatives, particularly grandmothers, as part of this research, and another small group of grandmothers from this community as part of an ongoing project on ageing and migration in 2013. The interviews confirmed that granny fostering is a well-established practice in this community. The number of grandparent/grandchild-only households in this community has increased from none in 2002 to 4.7 per cent of all households in 2008. Given the timing, the survey data also suggests that granny fostering is more common with the newer migration to Spain, than it was with the more long-standing migration to Argentina.

If we were to adopt a global care-chain perspective, the practice of granny fostering could be seen as one where grandmothers step in to substitute for their daughters' absence to care for their grandchildren. In the process, they also forgo the care that they might be receiving as older members of their families. However, in practice, this scenario only represents the experiences of those grandmothers whose health is deteriorating and are, in some ways, 'too old' to take on the additional responsibilities of looking after grandchildren. Many factors will impinge on this process: the age of the grandmother, her health, whether she is receiving an income, the strength of her social networks, and the age of the grandchildren.

Age is not chronological but is something that is experienced individually and is also socially determined (Vullnetari and King 2016). Therefore, the actual age is irrelevant to the ability to care for another person. What matters is how the migrant's parent feels and the state of her health. Many older women in this community receive their widows' pension so they do have some regular income coming in. This is important, given that remittances are often not regular, nor are they guaranteed. Whether grandmothers have other family members or neighbours they can rely on also matters, especially if they have a crisis.

Far from feeling abandoned and lonely, many older women in this community rejoice in being able to have an active role in the raising of their grandchildren. They are also very active in investing the remittances that their migrant children send them: buying plots, building houses, looking after investments, and often having their own businesses. As is common in many other lower-income countries, older people continue to work well into their old age and continue to contribute actively to their families.

Conclusion

The migration of Bolivians to Spain is quite similar to the migration examined in the Global Care Chains literature: for a larger proportion of women, care work is the most significant job at destination and separations are relatively long. However, there are also significant differences. As we have seen, many migrants eventually reunite their families, even in Spain. Combining work with family life was difficult for many migrants in Spain and impossible for those with 'live-in' work arrangements, but many migrants are able to get more flexible jobs and migrate with their partners or reunite their families. Those who were not able to do so often stayed for only a short period of time, relative to the migration of, for example, Filipinos to Canada. The work of Bolivians in Spain is also not as structured and controlled as those on the live-in programmes in Canada (Pratt 2005) or domestic workers in Singapore (Yeoh 2006) or Hong Kong (Constable 2007).

The exceptions to this greater flexibility and being able to combine family life with work abroad are single mothers, who stay abroad for longer periods of time and find it impossible to combine family life with work.

Migrants were better able to combine work and family life in their regional migration to Argentina. Some had trouble initially, which is linked to the garment work that many migrants undertake. The exploitative working conditions and the fact that many migrants start by living in the garment workshop makes it difficult to bring their children with them. However, most were able to leave the live-in working conditions for better-paid and more flexible jobs that allow them to have a family. As we have seen, some also set up their own workshop. The longitudinal and non-sectorial approach adopted in this research project allows for such analysis: the tracing of how migrants overcome difficult working conditions and resolve the problems they face by combining their migration trajectories with their personal lives. Gaining access to migrants through their community, as opposed to the sector they work in, also allows for insights on how migrants leave some types of work and obtain others. The inclusion of the 'return phase' also means that it is possible to advance a better understanding of the successes and also unsuccessful migration cycles, the cyclical nature of migration, and how important care considerations are for migrants' return, for both men and women. As we have seen, those who are not successful in combining their family life with their working lives often return home. There is often shame in this, but at least it is an option and they can always start again.

In conclusion, this chapter has argued for a broader perspective on social reproduction, one that includes regional South-South and Global South-North migration flows, including men and women migrants. This chapter has analysed how social reproduction and care influence migration flows and how they are in turn reconfigured through regional and South-North migration, and through time. Such an analysis gives a more varied and less bleak picture of the relationship between migration and social reproduction, one that goes beyond seeing the restructuring of social reproduction in transnational social fields as a 'care deficit' (Yarris 2017).

7 Intimacy

Intimacy and itinerant migration

As our understanding of migration evolves, so does our awareness of the multiple aspects of migrants' and stayers' lives that are shaped by migration. During the last decade, there has been an increased interest in intimacy and the ways in which migrants' intimate lives change through migration (Bloch 2017; Boehm 2011, 2012; Constable 1997, 2014; Walsh 2018; Faier 2009, 2011). Bloch (2017) argues that:

> the terms 'intimacy' and 'the intimate' helpfully demarcate an affective sense, one that is shaped by the forms of mobility men and women in this region have been engaging in since the end of the Soviet Union. The term 'intimacy' is inclusive enough to help bridge structural shifts facing people like labor migrants out of the former Soviet Union *and also* the personal, often emotional negotiations these same people are caught up in. (19, emphasis in the original)

Bloch draws on Wilson (2012) and uses the term 'intimacy' to blur the lines between migrants' personal and economic aspects of their lives and to encompass a range of relationships, including those "between parents and children, husbands and wives, temporary migrants and their close friends and boyfriends, the realm of domestic household space, and the sense of belonging that was lost with the end of the Soviet Union and the region's insertion into a global economy" (Bloch 2017: 19). Intimacy, therefore, relates to migrants' close and personal relationships, at the emotional level, which have been affected by increased geographical distance as a result of migration. At the same time, through migration, migrants forge new such relationships.

Most of the emerging literature on intimacy and migration focuses on migrants' personal and familial relations, albeit with explicit attempts to explain how structural processes, such as state policies and globalisation influence migrants (and stayers') intimate relationships. Boehm (2012), for example, is interested in how the US state is present in transnational

Mexicans' intimate relations. Through multi-sited ethnography over a long period of time, she explores how intimacy shapes migration and how, in turn, it is shaped by migration. She argues that: "This process is dialectical, one in which intimate selves and relations guide migrations, and wherein transnational movement is changing multiple subjectivities" (Boehm 2012: 11). Similarly, Bloch (2017) uses "'intimacy' because it allows us to avoid separating 'the economy' from 'the private'; thinking in terms of 'intimate economies' emphasizes how the lines between market/public space and private spaces are intertwined" (19). In both these interventions, the intention is to link macro-level processes with migrants' personal and intimate lives.

While the use of the term 'intimacy' to describe migrants' personal lives is relatively recent, research that deals with migrants' personal relationships and how these are affected by transnational migration is not new. Constable was one of the pioneers in this field, with her book *Romance on a Global Stage* on 'mail-order' marriages (Constable 2003). Parreñas also touched on intimate relations, but between parents and children, in her book *Children of Global Migration* (Parreñas 2005). More recently, Pratt wrote a moving account of the predicament that migrant mothers face in *Families Apart* (Pratt 2012). This research is also not confined to migration across borders. Research on indigeneity, for example, has also shown how internal migration influences migrants' and stayers' intimate lives (Canessa 2012).

Research on intimacy overlaps with that on emotion. Migrants' personal lives as well as the ways in which migration is often capitalised upon for political purposes are very closely connected with emotions and emotional responses to migration (Boehm 2011; Ahmed 2004; González-Fernández 2016). Some would see emotions as feelings and psychological states that might reside or emerge from the body. Others, however, would argue that emotions are "social and cultural practices" (Ahmed 2004: 9) shaped by the social, political, and economic structures within which the person experiencing the emotion resides. When migration is associated with suffering or desire, or when women joke about their partners' infidelities, these are not just personal feelings, but ones that reflect wider gendered structures of inequality (Boehm 2011).

In this chapter, I draw on these studies and my interviews to explore migrants' intimate lives and how these have changed through migration. I also take this discussion further and argue that migrants' working lives and the relationships established through paid work also often become part of their intimate lives, particularly in some of the jobs that women migrants take up, such as live-in elderly care or child-minding. When migrants spend the majority if not the whole day with the people they care for, as shown in the previous chapters, their working lives are also often embedded in a sense of intimacy, given the many hours migrants spend together with the people they care for, sometimes in very confined spaces.

Leaving children behind

Leaving your children when you know you are going away must be one of the most difficult decisions that migrants take, especially when they do not really know for how long or what their life will be like. I remember being struck when conducting the interviews by these stories of parents who leave their children to be able to fulfil their migration ambitions, especially those in Spain. At the time, my son was 12 years old. I have now re-listened to these stories with new insights, having a 15-month-old baby and finding it difficult to leave her at nursery just for the day. Mothers, but also some fathers, often have no words for the sadness they felt when they left their children to go to Spain or to Argentina. As Catalina said, it was *"algo inexplicable"* (San Fernando, 11th May 2009), something that cannot be explained or put into words. They often deal with this lack of expression by recounting the details of the day they left, or highlighting some detail about the process of separation that congeals their feelings. Sandra had tears in her eyes when she recalled the day she left:

> It was a Wednesday, 6th March, when we left, all the children crying. [Did they know? I asked] Of course they knew. The eldest understood. But the one who didn't understand was the middle one. He laughed, laughed, he wasn't reasoning but when he saw me go out with the luggage he realised [what was about to happen] and started shouting and he ran after the taxi, and like this, we left, a Wednesday evening. [And the little one?] He stayed, he was little (16 months), he didn't understand anything but the middle one, yes, he took it badly, and the eldest, both of them. (Cochabamba, 23rd May 2008)

Another interviewee suggested that she failed to grasp fully the consequences of leaving her children. Fernanda's husband was working in Santa Cruz, so she was on her own looking after her two children when she decided to go to Spain: "I left my children without thinking. I did my documents in a few days. There was nobody to look after my children. My mother was in Villazón, so I called her and she came that week, the day before I left. And I left them" (Cochabamba, 29th April 2008). Similarly, doña Josefa, who had been away for exactly two years, uses language that indicates she felt that she abandoned her children, despite the fact that she left them with their father: "when I abandoned them, I think, they must have been on their own" (Cochabamba, 5th May 2008).

The act of leaving their children is often frowned upon even by those women who left their own children behind. However, they try to create a distinction between their own actions and those of others, some of whom they see as going against their natural instincts. Fernanda, for example, who left her own son and daughter to go to Spain for eight months, talks of other women being *madres desnaturalizadas*, or denaturalised mothers, when they leave their children for three or four years (Cochabamba, 29th April 2008). The timing she indicates is quite interesting, as it is quite precise, and I would suggest, is probably related to the fact that the really good friend of mine who left her children to go to Spain is also her friend (they had travelled together) and left her children for two years. She might have specified 'three or four years' so as not to indirectly criticise the friend we have in common.

Josefina also said that leaving your children with somebody else leads to too many preoccupations: "when you leave them with somebody else, you always worry, you don't sleep, some days you don't eat, the worry kills you. That's why I want to take them with me [when I go back to Argentina]" (Cochabamba, 8th June 2002). Living separated from one's children imposes additional difficulties for settling into their new surroundings: "I couldn't get used to it. Because of my children, most of all, you can't leave them by themselves. I was remembering them every day. I just couldn't be" (Cochabamba, 8th June 2002). Valentina stayed in Buenos Aires only seven months because she was missing her children, but also because she had left them with their father, who decided to go to the Chapare region, leaving them on their own. A social worker who was at the time working in this neighbourhood was threatening to take them to an orphanage if Valentina did not come back. Her youngest did not want to see her when she came back.

For some the separation was an embodied act. Mothers in Spain talked about having to stop breastfeeding ahead of time as part of the preparations to go to Spain. Erika explained: "I left her when she was seven or eight months old. I had stopped breastfeeding three months earlier, because I was still breastfeeding her, and I left her like this" (San Fernando, 12th May 2009). Similarly, Catalina, who left her daughter when she was a bit older, also mentioned stopping breastfeeding as part of her preparations to go to Spain:

> Ay, it was sad, my mother-in-law came to say goodbye and my baby was little, she wasn't even two yet and I was still breastfeeding earlier 'but you're not going to your death, you are just going to work' they were saying to me. I was so sad to leave them. It was something that I cannot explain. I was crying all broken down and was remembering them and my mother was saying 'you are crying tonight but there [in Spain] you need to be like new, she said 'you are not going to your death, so go to your room and cry with your daughters and tomorrow you have to be like new'. It was really sad. (San Fernando, 11th May 2009)

Here Carolina also refers to having to be 'presentable' when crossing a border, when she mentions having to be 'like new' (see Chapter Four). Marisa also mentions that she cried when she travelled to Buenos Aires, leaving her children with their father. She cried and missed them. The youngest was 11 years old, so quite a bit older than the other children mentioned above. But even then, she mentions that "I left him when he was really small". He resented her, but she felt she had no other choice (Buenos Aires, 20th February 2003).

Although men in general were less likely to talk about the emotional aspect of leaving their children (or partners) behind, some did and went into quite a lot of detail about how they felt about leaving their children in Bolivia. Dario, for example, talked of the pain of being separated from his children and not wanting to repeat that experience. He was living with all four of his children at the time of the interview. When he first travelled to Buenos Aires, following his wife who had gone eight months beforehand, he left their three children in the care of his wife's mother (their youngest was born later, in Buenos Aires).

> She was working and I also had to find a job, but there was no work at the time. I was there for a month, two months, three months without work. 'Let's go back' I told her, 'let's go', I was also starting to worry about my children as well. I had left them with my mother-in-law and sisters-in-law. (Buenos Aires, 2nd March 2003)

He wanted to go back, but his *compadre* encouraged him to stay and work so that they would not go back empty-handed. So both he and his wife were working and regularly sending money to Bolivia for their three children. After about two years of being away, he received the news that his daughter was begging for milk, despite the fact that they were sending $300 to Bolivia every month. This is quite a large amount, as most migrants in Buenos Aires used to send between $50 and $100 per month.

> I started feeling for my children more, I mean, I started loving them more. I didn't know love, yes, I didn't know love. So I started realising that I love them, right? Only then I started knowing what love is. Well, and on one occasion, we were both working, we heard a rumour that our youngest had been seen asking for milk. We were sending money, $300 dollars. So there was a rumour that our youngest had been begging the neighbour for milk, with her bottle: 'Milk, milk', she was saying, it was a Saturday, my wife came home crying and said 'how can it be, we are sending money, why is she begging for milk?' She left her job. I cried. We both cried. We had money as well, we were working well. (Buenos Aires, 2nd March 2003)

So his wife travelled to Bolivia and brought their children to Buenos Aires, a journey that took two weeks at a time when there were no mobile phones. He had not seen them for two years, except for the youngest, who was born in Buenos Aires and then his wife took her to Bolivia. Others also felt the distance and missing their children led to a quicker return:

> When I was there with my family I was getting used to it, but this last trip, no. I was desperate especially for my children, I wasn't with my family, I used to call them, but it's not the same as when you're with your family, you share your daily life. We talked on the phone, but it wasn't the

same. One feels bad being in another country like this. (Cochabamba, 18th June 2002)

It is important to acknowledge this pain and suffering (Pratt 2009). Although the interviews were not geared to soliciting particularly painful episodes in migrants' lives, these emerged as part of the life story or testimony interview approach adopted in this research. Silencing this part of migrants' experiences would imply not respecting the trust that migrants placed in the interview process and in me as the interviewer. However, I also wanted to highlight the dangers of just reporting these experiences with the risk of representing these migrants' lives as 'different' and 'distant' to those of the writer (me) or the readers (you). The aim is therefore to place these experiences within a broader analysis of in/justice, to promote an understanding of migration, work and gendered regimes that combine to keep families separate, as will be shown in the sections that follow.

Consequences

The sadness and pain of leaving your children does not end with having to leave them. Those who have been reunited with their children after long periods of time are often painfully reminded of the prolonged absence by the children themselves. The children often do not recognise them as their mothers because during the absence they had become attached to others, usually their maternal grandmothers, who they also call 'mum'.

Daniela had a son when she was 15 years old and left him with her mum when he was seven years old to go to Buenos Aires. There she married and had two more children, but did not go back to Bolivia for eight years. Initially, she just got used to earning money and was spending it on herself but then, when she got married, she was not allowed to send money to her mother because her husband spent it all. She also did not have any money to visit her mother or her son. Eventually, she left her husband and was able to travel to Bolivia when her mother was dying and her son was in his final year of secondary school. She arrived the day before her mum died and recalled her son saying to his grandmother:

'How are you going to die, Mummy, don't go, who is going to accompany me when I graduate from high school?' She clarified: He doesn't call me 'mum'. He calls me by my name. He doesn't call me mum. [...] He calls my brother 'dad'. Yes, this is how it is. (Buenos Aires, 15th February 2003)

Erika asked her aunt to bring her daughter to Spain when she was two years old. She thought that her daughter was not going to accept her. She recalls how at the airport her daughter only spoke to her dad, because he stayed in Bolivia for a while longer before joining Erika in Spain. She hugged him and he had to tell her that Erika was her mum. She said: "She looked at me and smiled, but she didn't want to come close, she was treating my aunt as if she were her mum, anything she wanted, she asked her, not me. [...] After a week, she called me 'mum'", although she was still asking for her 'other mum', her grandmother. It was customary to call grandmothers *mamá* in the mining town. Erika in fact explains that to call a grandmother 'granny' is considered impolite. In many families, the grandmother is often called 'mum', or *mamá grande*. "In the mining town, we always call grandmothers 'mum' because to call them 'grandmother' is as if you were disrespectful" (San Fernando, 12th May 2009).

Mothers who left their children when they were very small are often called 'aunty' by their own children once they are reunited. My friend who I stayed with during fieldwork had only been back for a few weeks when I arrived in Bolivia in 2008 and her youngest son, then three, used to call her 'aunty' and called his grandmother 'mum'. I was 'aunty number two' (fieldwork notes, 18th April 2008).

Those who went back to Bolivia after long absences found their children different. Patricia, who left her son for two years, said that he was not the same as the one she had left when he was seven years old. Patricia and her husband travelled to Spain in 2005 and left their seven-year-old son with her mother. "*Me esperaba bien cambiadito*", she said: "He had changed a lot during my absence". She complained that he did not really respond to her, i.e. he does not listen to her, "*ni en palabras ni en ...*", "not when she talks to him, nor..." leaving the second part of the sentence unexplained. She had been back for almost two years when I spoke to her, but she said that "he's not the same as when I left him" (Cochabamba, 20th May 2008).

Some were rejected even after relatively short absences, like Valentina, introduced above. She said that her youngest did not want to see her when she went back after seven months of being in Buenos Aires:

> They cried so much when I arrived. My youngest, he didn't want to get close to me. He only looked at me from afar. I told him: 'Come here, baby' but he didn't want to. He didn't want to. He was scared, I don't know. He didn't want to get close to me. When I got close, I kissed him and he just cried. (Cochabamba, 8th June 2002)

Fathers also noted some differences in the ways in which their children related to them after they had been away for some time. Daniel, who had been in Buenos Aires on a number of occasions between 1994 and 2002, mentioned that his youngest did not want to sleep in her parents' bed since he had been back (Cochabamba, 9th June 2002).

Many, like Sandra, complained that children "get lost", "*se echan a perder*" (Patricia), or "*se descarrillan*" and then gave examples of other people's children, who "lost their way". There was a sense that there is a fine balance to be struck between working towards achieving one's aim of, for example, saving $20,000 to finish building a house, and returning home fast enough to be able to still catch your children before they stray off the right track.

> ... In terms of children's education, terrible. It's so sad. Such a big problem. Clearly, migrating is great. It brings you riches, fortunes. But on the other hand, it destroys the children ... unless you come back in time and you can save them, but many don't come back in time. It's more, they disappear. (Cochabamba, 23rd May 2008)

Community leaders, nuns and teachers expressed similar views about migration and the problems that in their view it generated within the community. However, it is interesting that migrants themselves also adopt very similar points of view. Research carried out in Ecuador has similarly highlighted how concerned migrants are about who is going to discipline their children during their absence (Boccagni 2013).

Migration and domestic violence

As already mentioned, domestic violence is quite common in this community. This is not to say that there is a high incidence of domestic violence. I do not have the data to support this statement, but the way in which people talk about domestic violence indicates that this is not an exceptional occurrence. People might complain about excessive domestic violence, taking it beyond an acceptable level of domestic violence where there is such use of force and damage done. However, people do not usually complain about domestic violence per se. This became clear in a discussion that I recorded in my notes from fieldwork in Buenos Aires:

> There have been a number of very vivid recollections of cases of domestic violence today. In the first one Evelyn told me how her brother-in-law was once beating his partner, Evelyn's sister, after coming back from a party. He was pulling her by the hair and she was bleeding when he took the telly and wanted to break it over her head. Evelyn and her sister-in-law were both in the room but didn't dare intervene because he had pushed his sister-in-law across the room and out of the way. When they saw him with the telly, her sister-in-law told Evelyn to get hold of the telly. We burst out laughing as it seemed that Evelyn was more worried for the telly than for her bleeding sister on the floor. While we were expecting her to say 'I was worried about my sister', she said 'I was worried for the telly'.
>
> The second one involved Catalina and Dominico. They said that Dominico beats his partner Catalina a lot in Bolivia. One day when he was drunk he beat her so much that my friend and her mum hardly recognised her when she came to them crying. Her face was all swollen and they told her to leave him and asked why she was staying with him.

He spent a few hours at the police station and then came looking for her and managed to convince her to forgive him. (Buenos Aires, 28th January 2003)

Domestic violence was also commonplace in the mining town from where most of my interviewees originate. As mentioned in Chapter Three, some recalled how in the mining town it was almost expected that men would beat their wives "so that they would behave", as don Daniel argued. Others have also shown that domestic violence is common in indigenous communities, but that it is to some extent regulated through the woman's family relations (Canessa 2012). Should the violence be too severe, or too frequent, then her brothers are expected to intervene to mitigate the situation and protect the victim.

Doña Rubina, whom I interviewed in Madrid, recounted her own experience of years of suffering at the hands of her husband, whom she was forced to marry after he had abducted her when she was 15 years old. She said she did not 'live well'. One night, her brother left her with one of his daughters, because her husband had threatened her. She had her nine-month-old baby on her back when her husband hit her from the back. She narrated how he pulled her hair and she fell back onto her baby. He was beating her up while her niece went to get her brother. She was covered in blood, lost teeth and had four stitches on her lip and mouth. Her brother came to her rescue and took her to hospital. She separated from her husband after that (Madrid, 8th August 2008). As others have argued, brothers are essential for making sure that any domestic violence remains within acceptable boundaries. Clearly sometimes the intervention comes too late or is not effective and women suffer both physical and psychological damage as a result of the beatings. This becomes more difficult when women migrate, because they are far away from their sources of support. Sometimes brothers are the perpetrators, as was the case with one of my interviewees (see Chapter Four). In cases where women experienced domestic violence in Argentina or in Spain, they generally sought out help among people who were not from their transnational community. In the case of the woman who was beaten by her brother, the person who helped her was an Argentinian friend, who housed her for a few months after this happened.

Although this might just be a coincidence, the discussion about social networks in Chapter Four would also suggest that women would not find much support from within their communities. Women were less likely to travel with friends and have 'community-wide' social networks. They tended to travel alone and find accommodation in hostels, or with family. It therefore makes sense that in times of crises their access to extra-family, community-wide social networks would also be quite weak.

Women who are victims of domestic violence while they are abroad are faced with a different institutional context. Their migration status might also prevent them from seeking help and support from the authorities of their country of residence. In Buenos Aires, through interviews conducted for another (but related) project, we found that migrant organisations were claiming to be providing support for victims of domestic violence. However, they were doing this in a system parallel to that of the relevant national authorities. Their tendency was to seek reconciliation between the couple involved and they tended not to report the events to the national authorities (Bastia 2018).

High levels of domestic violence are often at the root of women's migration trajectories. Doña Paolina, whose story was already recounted in Chapter Four, left the mining town and ran away from her husband who used to beat her severely. She crossed the border into Argentina with her two children in the 1960s and continues to live in Buenos Aires. Evelyn's sister, who was beaten until she was on the floor bleeding while her husband lifted the television threatening to smash it over her head, left for Spain after returning to Bolivia and remained there. People commented that she was reluctant to return to a violent relationship, as were many others.

However, not every woman who had a violent relationship decided to stay abroad. Others migrated as a signal to their partners that they were not happy with the relationship. Ana suspected that her husband had been unfaithful while they were in Buenos Aires. He also used to beat her a lot. She left for Spain, leaving their daughter with her sister-in-law, despite his opposition. He followed her a few months later and at the time of the interview they were together. She said he now understood her better and made more of an effort to have a better relationship (Madrid, 13th May 2009).

For those living with domestic violence, migration was also about achieving economic independence, not necessarily by choice but because, aside from being violent, their partners and husbands did not contribute to the financial maintenance of their families. They used migration to earn a livelihood for themselves and to support their children, while their husbands were often 'lost' for months at a time. However, they stopped short of formally separating from their husbands or seeking another partner, because, as one now grown-up daughter explained in relation to her mother: 'men don't respect single women' (29th July 2009).

"In Spain men kill their partners, but in Bolivia they only beat them": comparative perspectives on domestic violence

Migration clearly presents an opportunity for migrants to compare and contrast different practices, across two, or more, social settings. In relation to domestic violence, more than one interviewee commented that levels of domestic violence in Spain were excessive, as compared to what they were used to in Bolivia. Sonia, for example, was quick to acknowledge that domestic violence was endemic in Bolivia. However, bringing a comparative aspect to her analysis, she was surprised about a specific case that was shown on the Spanish television, where a man stabbed his partner four times. She said:

that they kill them, here I see that it's normal, they showed on the telly a case where he had stabbed her four times, *es una barbaridad*, [it's shocking], but there [in Bolivia] it's not like this, there is violence, but with knocks, as if they were hitting a sack of potatoes. (San Fernando, 29th July 2009)

Later in the interview, she linked this high level of violence to two things: the legislation that protects women in cases of separation and also the fact that women hold their ground and 'answer back' in discussions with their partners. In relation to legislation, she felt that women are given too many advantages in cases of separation, such as being allowed to keep the house they live in while their husbands have to continue paying the mortgage. In her view, this made men resentful towards women. Moreover, Spanish women were perceived as being too arrogant in their personal relationships, too individualistic, as also mentioned by another interviewee, who would answer back in an argument, compared to Bolivian women who might play a more submissive role and cease more ground to their partners. Spanish women are therefore seen as being too autonomous, in the sense discussed in Chapter Two, where they are seen as promoting their own well-being within a context of separatedness, instead of seeing their well-being in the context of the families they live in. She argued that women in Bolivia used this strategy so as not to escalate an argument and decrease the chances of being beaten, or of being beaten severely. In their view, Spanish women, by 'answering back' and not ceding ground, unnecessarily wind up their partners and, as a result, the level of violence they receive is sometimes lethal. While these attitudes represent a view of women's roles that is more firmly embedded within the social relations that surround them, they nevertheless also stem from a misogynist vision in which women are blamed for the violence that they receive.

At the time of my fieldwork, the Spanish government was running a national awareness-raising campaign about domestic violence, part of an international initiative called *Maltratozero* – or zero tolerance to violence (see Image 7.1) The campaign has probably contributed to Bolivian women perceiving domestic violence as being a greater problem in Spain. However, the national statistics on domestic violence do not support these arguments. Bolivia, in fact, heads the regional statistics as the country with the highest incidence of physical violence in Latin America and the second highest for sexual violence (Requena Gonzáles 2017).

It is estimated that in Bolivia over one million women over the age of 15 suffer or have suffered some type of physical violence in their intimate relationships, including having sexual relations against their will, out of a total of 3,321,781 women who are currently in a relationship or have been in the past. In 2016, over 200,000 of these sought help from an institution and 141,245 presented formal charges (INE 2016). In Spain, a much larger country, the number of formal charges for domestic violence in the same year was very similar, at 143,535 (Secretaria de Estado de Igualdad 2016). In 2015, there

Image 7.1 Domestic violence campaign, Spain

were 56 fatal victims of domestic violence in Spain, which had a population of 46.4 million, and in Bolivia 68 cases of femicides, *feminicidios*, a broader category, but again, in a much smaller country of, at the time, 10.8 million (El Mundo 2015; INE 2015).

It is clear, therefore, that despite interviewees' perceptions that domestic violence was more severe in Spain, the relative number of femicides in Bolivia was over four times higher than the number of deaths as a result of domestic violence in Spain. Their comments, however, point to normative understandings of how women should behave in situations of violence and their expectations that women should not provoke their partners by answering back. There is an expectation that women should play down any conflict by remaining submissive and ceding ground to their partners so as not to escalate violence, or the threat of violence towards them. This is an important point, particularly when women are away from their extended families and usual networks of support, who are therefore unable to intervene in situations of violence.

Migration as sexual freedom

Migration is closely related to greater sexual freedom both in terms of the migrants themselves, but also for those who stay behind. Most interviewees who talked about sexuality linked migration to greater sexual availability and sexual freedom. A common saying at the time of my fieldwork in Buenos Aires was: *"aquí todos son solteros"*, or "here everyone is single", referring to the perception that when somebody travels to Buenos Aires, it is assumed that they will be sexually available, regardless of whether they left a partner in Bolivia.

Some, like Angela, felt that couples will separate if they do not travel together (Algeciras 4th August 2009), while others, like Sandra, were of the opinion that even if they travelled together, they would separate anyway because they find new partners (Cochabamba, 23rd May 2008). Sandra blames this on women and partly on the fact that couples are not allowed to beat each other (!):

> When couples travel together, they either tear each other's eyes out, or they fight, but over there [in Spain], they are not allowed to beat each other up, because the police intervene straight away, they end up separating... They start going out with each other's partners, or one fancies another person... Of 100 people, 70 per cent end up separating. Those who remain are those who work *de interna* or those who work with their husbands. But those who work elsewhere, *chau*, they separate. (Cochabamba, 23rd May 2008)

Sandra therefore understands domestic violence as a balancing act aimed at encouraging stability and equilibrium within a relationship. If couples are not allowed to 'beat each other up', their relationship is likely to suffer and they end up separating and seeking other partners. For others, like Marta, migration leads to greater promiscuity among migrants when they migrate without their partners, but also among their teenage children, who become precocious and awaken sexually before their time. In part, she blames this on their adopting European behaviour:

> There is a moment when things are not like this because the other person comes back with their own or simply finds another person and well, they've given themselves to the European lifestyle. Because life in Europe is like this, for them the family and the marriage isn't important. For many, it is but for the great majority it isn't. That's what I have seen. They marry and after two or three years, if that, they separate. [...] And the Bolivian always adopts bad habits first, not the good ones. That is the worst. For example, over there, cohabiting with somebody is common, normal, for the Spanish or the European [...] Instead, here it isn't, there are still those taboos, although you may not want to recognise it [...] all

these habits arrived with migration and they became normalised. (Cochabamba, 16th May 2008)

This is a striking statement, considering cohabitation is commonly practiced in rural areas as well as in the mining town. However, the statement indicates precisely the opposite – how middle-class expectations of marriage before cohabitation are being adopted in peri-urban neighbourhoods with high levels of migration, but being attributed to transnational migration.

Here it is also important to note that separations and extramarital relations are common also in couples where nobody migrates or among migrants, before they migrate. Therefore, a lot of the talk of migration making intimate relations worse should be put into this wider context of how common these practices are in couples where there is no migration involved.

Others have had a more direct experience of these processes. Laura, for example, migrated to Buenos Aires during the 1990s when she was still in her teens, following her father who had been there for a longer time. Her parents separated halfway through his stay abroad and she was left in charge of her siblings from the age of 15. While Laura was in Buenos Aires her youngest sister started cohabiting with her boyfriend. According to Laura, this happened because she felt lonely. At the time of the interview, she was very worried about her sister, especially because the man she was living with was known to have been abusive towards his previous partner (Cochabamba, 30th May 2002).

Laura, as well as others, commented on how there was often an assumption that migrants are available for extramarital relations, if single but also when separated from their usual partner as a result of migration. In the previous chapter, Javier already indicated the importance of keeping the family together given that he felt that intimacy was lost through distance (Buenos Aires, 31st May 2008). In his first interview in 2003, he also talked about men and women needing each other and that both men and women pretend to be single when they first arrive in Buenos Aires – "*todo el mundo es soltero*", "everyone is single", which leads to family ruptures.

Migrants sometimes blame the sleeping arrangements in garment workshops for promoting greater promiscuity. Men and women usually sleep separately but close by, and interviewees commented that men harass young women in garment workshops. Doña Valentina, for example, explained that in the garment workshop where she worked there were only two rooms, one for men, and one for women. There were more than seven people sleeping in each room, in bunk beds for three people each. Men used to harass women, but this did not bother her too much because they mostly harassed younger, single women (Cochabamba 8th June 2002).

Some migrants were reluctant to migrate alone because they assumed that their relationship with their partner would suffer as a result. They were therefore prepared to earn less, have fewer savings, or pay more for their trips to maintain their relationships. However, as indicated above, only a few went to the extent that Sonia did, travelling to Spain with her whole family, in her case, with her husband and two children, because of the difficulties involved (San Fernando, 29th July 2009).

Intimacy at work

As mentioned at the beginning of this chapter, the migration scholarship has started looking at intimacy and the ways in which migrants' intimate relations change through the migration process. Work is also one of the spheres where migrants may forge intimate relations. As already mentioned, migrants often work in personalised occupations, in which they provide personal services related to care work. While some of these end up being abusive, some migrants also build positive intimate relationships with the people they care for.

Only a small minority of migrants worked with children. But among those who did, there was a strong resistance to being on call 24 hours and, in their view, replacing the emotional and physical care that mothers are supposed to give. Marta had worked as a childminder for a period of time while she was in Spain. She was asked to work 14 hours a day but was living with her employers, who were also expecting her to get up during the night whenever their baby cried: "She wanted me to get up when the baby cried, as if I were his mother! [...] That's what I mean, they see us as robots that are going to produce, produce, produce..." (Cochabamba, 16th May 2008). This was in the context of a broader critique, in which she felt that migrants are not treated as people by their employers: "They imposed their timetable, I had to eat what they ate, even though I didn't like it. For them, they never took you into account as a person [...] They don't have that feeling that we are also people and should be treated as such. No. For them, we are there to serve them and that's it" (Cochabamba, 16th May 2008). In this case, Marta resented the intimacy that she felt was being imposed by her employers. Others, on the other hand, built positive relationships with the people they cared for.

Despite the long days and the undefined list of tasks these carers are responsible for, some of the interviewees cared deeply for the person they were looking after. In fact, some suggested that they were 'like daughters' to the women they were looking after, that is, that they provided the people they looked after with the love and support that a daughter would provide to her elderly mother (or father) in an idealised world and according to the moral expectations implied in Bolivian culture. The woman that Fernanda was looking after was 88 years old, but this job lasted for only four months because the person she was looking after died. In recalling her experience she invokes family relations to make sense of the relationship she had with the woman she was looking after, calling her 'granny' and positioning herself as a daughter.

My granny died. I looked after her in the hospital for a month, because she didn't have any children. I took her to the hospital, I did everything, it was as if I had been her daughter, I was looking after her, she had a scan, radiographies, I took care of everything, because the granddad [the husband] couldn't do anything, he was also 88 years old, he couldn't do anything. He had a daughter […] but she had a family, children, a husband, she couldn't… The lady I was looking after was in hospital, ill, because she wasn't feeling well, she had stomach pains, and her guts had burst. The doctors were only waiting for her to die. I asked them: 'When is she going to get better? Please operate her.' And the doctor would tell me: 'Miss, you know what, are you the main carer?' 'Yes, she doesn't have anyone else.' 'OK, I will tell you the truth. The lady is going to die. We are only waiting for her to die.' He told only me. I couldn't believe it. And I looked after her during the day and during the night. Before dying she complained a lot, she said: 'Ay, *hijita* [daughter, endearing] please look after me, look after me,' because she wouldn't let me sleep at night, she wouldn't let me sleep, she would shout: 'Take me out of here!' I suffered all that because I went to work. Then she died, we buried her. (Cochabamba, 29th April 2008)

In this quote, the responsibility that Fernanda felt for the elderly woman she was looking after is clearly placed within the realm of work, albeit with an understanding that work implies suffering. However, it is also clear that responsibility in this case is much more than just the daily care that might be required to keep an elderly person well and comfortable. Fernanda placed herself and was also placed by the doctor and the person she looked after within the realm of familial relations. She was given confidential information about the care that the elderly person was receiving in the final stages of her life, information which would generally be restricted to family members. The elderly person's death meant that Fernanda was also jobless, but she does not emphasise this part of the story.

This quote also reflects cultural expectations of care in Bolivia, where looking after your elderly mother is a moral obligation. Transferring these expectations to her role as elderly carer, Fernanda therefore positioned herself in the role of daughter and cared for her granny anyway, whether this was part of her job or not, as when she includes burying her 'granny' in her account.

This sense of responsibility expressed by elderly carers in some cases translated to migrant women's staying in their jobs even when they were offered better paid opportunities elsewhere. This implied a longer separation from their children. For example, Zenaide was born in the late 1960s and was a single mother whose daughter was 14 years old when she decided to go to Spain. She found work looking after an elderly woman who died two months later. She then found another job caring for a 92-year-old woman. She was paid 700 euros per month as '*externa*'. She was offered a better-paid job: 1,000 euros per month as live-in, but she did not take it, despite the significant savings the move would entail. She explained: "I got used to the granny" (Algeciras, 2nd August 2009). In this case, it is clear that Zenaide did not see her care work as a mere responsibility attached to a payment. Zenaide accepted to continue in a lower-paying job, which also delayed her own return to Bolivia and reunion with her own daughter so that she could continue looking after the 92-year-old woman.

Although Zenaide does not make this comparison, other migrants had a set target amount of savings they wanted to achieve. They aimed for the highest-paying job to maximise their earnings which would help them achieve their savings target as soon as possible to minimise the time they had to stay abroad and therefore separated from their children (see next section). While experiencing different types of harm (see below), many migrant carers made sure that their 'grannies' did not remain alone when they had to leave their jobs, either because of pressing issues back home or because they could not take the pressure any longer. These decisions are often overlooked in studies of care because they focus on those who are still employed in the sector. Fernanda, for example, found it very difficult to cope in her last job, but she also made sure that the lady was not left on her own: "But before coming back, I left another Bolivian there with them" (Cochabamba, 29th April 2008).

Sometimes the emotional responsibility they felt towards the people they looked after prompted them to continue working even when they would have preferred not to. Rosa, who was in her early thirties, returned early to work after a maternity break because the elderly woman she was looking after remembered her and was used to having her and her sister as her main carers: "Now I went back to work because, well, the granny got used to us, she has Alzheimer's disease, but she hadn't forgotten me and my sister" (San Fernando, 27th July 2009). These examples show how migrant workers felt responsible for the 'grannies' they looked after and experienced this as much more than a paid job (England and Dyck 2012). Despite the fact that the ways in which they made sense of their relationship kept them in a position of disadvantage – e.g. they did not contest low wages or seek a better job elsewhere – migrant elderly care workers did not always perceive these conditions in a negative way. They usually see their work in Spain as time-limited and do not have the expectation of having the same rights and protections as Spanish workers, which is common with migrant care workers in other countries too (Datta et al. 2010; Näre 2011).

These choices are not simply emotional but also a result of the conditions these migrants face, such as the visa for Bolivians introduced in 2007, which implies that it would be very difficult, if not impossible, to return to Spain in the foreseeable future. The cost of travel to Spain often makes a temporary return to Bolivia prohibitively expensive. Moreover, as mentioned above, the traditional countries of emigration, the US and Argentina, had lost their appeal due to new migration barriers for the former and financial instability for the latter. Carers also put up with difficult working conditions because they see their work abroad as temporary and they endure it with the aim of achieving their migration goals, e.g. a specific amount of savings needed to complete a project, such as building a house (Bastia 2012). Bolivians were the last among Latin American migrants to arrive in Spain and therefore faced more competition in the labour niches they were seeking to enter. Moreover, few were able to benefit from the amnesty in 2005 because they did not qualify for the residency required to apply for the amnesty. The debts that they often incurred at home in order to finance their migration also weighed heavily on their decisions to remain and put up with long working hours.

While the context is useful in understanding the choices these women made in terms of staying in Spain, it does not explain the relationship that these elderly care workers had with the people they looked after, and the sense of responsibility they felt towards them. Therefore, despite this broader political economy context, which helps understand the constraints that many migrant elderly carers have to negotiate, their acting responsibly does suggest that they practise an ethic of care (Lawson 2007) towards distant others by taking on a much larger role than just the tasks that are associated with keeping an elderly person well and comfortable. They feel responsible for their well-being. The working conditions described above also raise questions regarding the carers' difficulties in looking after their own parents and children, the responsibilities that are denied to them.

Conclusion

Intimate relations with partners, children, and the people migrants care for at work shape how migrants experience migration. These intimate relations bring global inequalities and unequal structures of opportunity at both origin and destination to bear on migrants' most personal relationships, what they experience, and what they feel about these changes. Migration can be painful and sad, but migrants also forge new intimate relationships while seeking work abroad.

Looking into migrants' intimate lives also sheds some light on their views on gender inequality, for example, in relation to domestic violence. Migrants' experience of domestic violence is clearly not just gendered but also influenced by their ethnic background and their class position. We have seen how some do not condemn domestic violence but propose a relativistic view of domestic violence that ends up justifying its practice, as long as it is not excessive.

8 Conclusion

Migration and social transformation

Migration is intrinsically linked to social transformation. As we have seen, migrants from this community started migrating to Buenos Aires because of a 'failed' internal migration. Their international migration was therefore directly linked to the economic restructuring and changes at the global level that led them to lose their main source of livelihood: mining. After moving to Cochabamba, they failed to secure a firm positioning in the urban labour market so many decided to try their luck in Buenos Aires. Then, after the Argentinian crisis, their migrations shifted towards Spain. These international migration journeys were therefore a direct consequence of the economic and social transformation that was taking place in Bolivia during the 1980s and those in Argentina during the 1990s and early 2000s.

At the same time, migration has also contributed to significant social transformation in this community of origin, a neighbourhood in peri-urban Cochabamba. From a relatively strict male breadwinning ideology, women started integrating themselves in the world of paid work outside the home in greater numbers, including in their search for work in Argentina. By doing this, they were directly contributing to the process of feminisation of Bolivian migration in Argentina. Capitalising upon their regional migration experience, these women then took the lead in the then emerging migration to Spain. Through this process, women went from being 'housewives' and, at most, 'helping' their husbands with petty trading in the mining town to actually being the breadwinners of their families. They started financing the construction of their family homes as well as putting down the capital for family enterprises. The question is: have women benefitted in this process? If so, what have they gained? To what extent can their migrations be seen as emancipatory?

To answer these questions, the empirical chapters dealt with four main areas that are fundamental for understanding gendered migration processes: mobility and social networks, work, care and social reproduction, and intimacy.

From gendered social networks to intimate itinerant migrations

We have seen that men and women access gendered social networks when organising their migration journeys. The evidence indicates that between the two survey periods (2002 and 2008) there has been an increase in both men and women travelling alone. This trend is supported by the literature on migration and social networks, which shows that as a community becomes more enmeshed in chain migration, community-wide social networks become less essential for new migrants (Winters et al. 2001). Once migration has become established, family networks are generally enough. The transnational community under consideration in this book underwent similar changes. From an initial mostly male-led migration, making use of community-wide, friendship-based social networks to facilitate migration to Argentina, the modality of migration changed to one that was mostly women-led and family-based, to various cities in Spain. This new modality of social-network use also went hand in hand with a different geography of migration. In Buenos Aires, members of this community found accommodation and often work in three informal settlements. On the other hand, in Spain, migrants from this same community scattered from the very northern tip of Spain to the Canary Islands. The women whom I interviewed were able to make the most of these opportunities and used the social networks available to them to help them find accommodation and work. When their origin-based social networks were closed to them, they diversified and sought support from Argentinian neighbours. This enabled them to enter the destination labour markets and take advantage of the opportunities available to them.

Migrant women from this community are firmly embedded in the world of work. The main reason for their migration was paid work, to a greater extent than men. This is because, over time, women's work has become fundamental for the financial sustenance of these ex-miners' households. Their integration into the world of paid work was progressive, from petty trading and 'helping' their husbands in the mining town, through a significant engagement in trading in Cochabamba and then garment work in Buenos Aires, where they generally worked on a par with their partners, but for half the money. The migration to Spain, though, opened the door for them to accumulate significant savings over relatively short periods of time, albeit for higher risks and for virtually no choice in the type of job they did. Once in Spain, all they could do was elderly care work, usually live-in, with some exceptions for those who did domestic work by the hour or entered atypical occupations for women, such as construction and restoration work, but this was just one case. However, in order to get there, they had to borrow money and overcome the risk of being deported upon arrival to Spain. Most had to live with their undocumented status. The fact that only one interviewed elderly care worker managed to regularise her stay and thereby her work by 2009, after Bolivian migration to Spain had already peaked, benefitting from all the rights and security that came with this, illustrates the daily difficulties that this group of migrant women had to endure.

Seeking work abroad also often involved separation from loved ones. Chapter Six showed that migration often precluded family life. The work, gender, and migration regimes made it virtually impossible for migrants to have a family life and bring their children with them, particularly to Spain but also to Buenos Aires. Although some have been successful in overcoming these difficulties, especially following an initial period of adjustment, many were unable to reconcile their parental responsibilities with those imposed by the jobs they were able to access. There are costs, yes, but these are difficult to quantify, especially when these are framed in contrast to women's greater autonomy.

One would expect that reconciling migration with family life would be more difficult with global, South-North migrations, but recent migrants in Buenos Aires faced similar challenges to those who migrated to Spain: high rents, inadequate accommodation, low wages, and long working hours. However, most regional migrants in Buenos Aires were able to reunite their families and enjoy family life in Buenos Aires. For those who were unable to do so, temporary migrations and earlier returns were part of the strategies they implemented. Migration theories, therefore, need to be better able to integrate return, actual or potential, as an integral part of migration processes.

This chapter also showed how care and social reproduction are flexible and adaptable in the context of regional and global migrations. In line with other emerging research from the Latin American region, I argued that instead of conceptualising care as a 'deficit', as suggested by the global care chains literature, it would be more useful to understand how traditional structures of care change to accommodate the new reality of migration (Herrera forthcoming; Yarris 2017). If one adopts a historical approach, one realises that the granny fostering that most migrants rely on to provide care for their children in their absence, is a long-established practice that pre-dates transnational regional and global migrations. People already relied on granny fostering in the mining town. Adopting a broader and historically situated analysis, therefore, arrives at a less pessimistic view of the changing care practices resulting from migration.

Taking another step towards better understanding the intimate experience of migration, the last chapter explored intimacy and its role in itinerant migration. Building on recent work on intimate geographies and intimate migrations, it showed how migration is fraught with sadness, which sometimes takes an embodied form, yet at the same time, this personal suffering also represents wider social and economic structures of inequality (Walsh 2018; Boehm 2012). Some of these are related to the difficulties of reconciling migration and family life, discussed in the previous chapter. However, the sadness experienced as a result of separations with loved ones is often accepted by the migrants themselves. This 'cost' has to be understood in relation to the sadness of, for example, not seeing one's child through to education or through an operation because without migration, the child would not have had these options. These choices or migration itself, despite being 'voluntary', is far from it. Remaining in Bolivia when there is an opportunity for better-paid work – or just for work itself – elsewhere, is just not an option. Clearly, children are left because there is still a family structure, either nuclear or extended, that is there to support these projects.

Other examples are related more directly to gender inequalities and gender-based violence that many women experience. Some of the stories included in this book show how migration is, for some women, the only escape route from violent relationships. The women who leave their violent partners, will sometimes be ostracised by the community itself. Many of the cases that I heard of were of women for whom it was difficult to obtain contact details, because they lived in cities other than the ones where I was carrying out fieldwork and they had lost touch with members of this community. Women who leave their partners often had to also give up their children, given that their fathers would not allow contact with them. This takes us back to the issue of choice: to what extent are these women actually choosing to migrate and leave their children behind, when they are only seeking a safe place away from their violent partners?

Changes in gender roles and awareness of gender inequality

There is no doubt that migration entailed changes in gender roles. Women went from being housewives, mainly staying at home to look after multiple children and being economically dependent on their husbands, to being the family breadwinners. Because of the broader changes that shook this community, today there is an acceptance and an expectation that both men and women would work. This also extends to an acceptance that if better-paid jobs are available to women, they will take the highest-earning job. However, the danger here is that women just end up doing more, without a reconfiguration of their personal relations towards greater gender equality.

Migration did bring about some additional changes in the sexual division of labour. The survey showed that in migrant households there is a redistribution of housework away from women and children, towards men and other female adults, both from the family and brought in to help through paid work. However, women tend to downplay their new roles and their achievements. They also display a lot of ambivalence towards the new configuration of gender roles.

There is, however, definitely some change in women's and men's awareness of gender inequality. Migration allows both men and women to become aware of the socially constructed nature of gender relations. They realise that in the places they migrate to, gender relations are organised in a different way. In some instances, this has led women, in particular, but also some men, to adopt a language of gender equality. However, instances of men or women who went against traditional norms of understanding femininity or masculinity were few and far between. For example, the notion that 'women serve men' was not contested by interviewees, not even those who were vociferous against domestic violence. Moreover, in intimate relations, women continue to face violence. As I have shown, domestic violence is still accepted. Even those who understand that domestic violence is not as acceptable in Spain as it is in Bolivia, employ a relativistic discourse that ends up justifying this practice 'at home'.

Is migration emancipatory?

Therefore, what do women gain in this process? Most women, as we have seen, gain better access to jobs. They are also able to access better-paid jobs. Their income-earning power is therefore higher. Sometimes, they take on new roles, such as, becoming the family breadwinners. There has been some renegotiation of household tasks, so there are some changes in sexual division of labour, away from women and from children, towards men and other women (family members or paid workers). However, to what extent do these changes 'stick'?

Evidence suggests that these changes are largely temporary as, for example, when migrants talk of this being a 'moment of matriarchy'. Many women return home after having been abroad to find that there are piles of washing to do. While abroad, they have even more limited choice in terms of the jobs they can access than in their home labour market, and this reduces progressively, from being more restricted in Argentina to having no choice in Spain. The type of work that they carry out is generally poorly paid, informal, and often exploitative, lacking the most basic labour rights and job security. Therefore, the 'higher earning power' is such but only in relation to what they can achieve in Bolivia. It is also only temporary and lasts as long as their migration.

We cannot therefore call this migration emancipatory because of the types of jobs that women carry out, the limited or lack of choice they have, and the temporary nature of the changes that take place. While women do migrate autonomously and often make the decision to migrate, even in contexts of 'joint' or 'male-led' migration, this does not necessarily imply that migration is also empowering. While women do participate in the decision-making and are sometimes the sole decision-maker, they take these decisions in contexts of vulnerability, increased precarity, having to shoulder greater responsibilities for the financial upkeep of their families where the only types of jobs they have access to are insecure and generally poorly-paid.

There is also no direct challenge to patriarchal norms. Women still generally serve men. Nor is there a challenge to patriarchal structures: the nuclear, heterosexual family is still the preferred family form for most interviewees.

Having started this project with very firm feminist objectives, the findings have been somewhat disappointing. Migration has definitely contributed to economic and material improvements. However, in terms of gender equality, the only clear evidence is that women tend to trade whatever gender gains they might have accrued (through a better economic positioning within the household) for class gains, as they see migration generally in terms of providing the means for upward social mobility.

Migration is not a project of emancipation. So if gender gains and feminist social transformation take place, this is due to feminist movements pushing the government for greater accountability and changes in legislation and to actions against gender-based discrimination and violence against women. Migration can be part of this process, in terms of migrants acquiring greater awareness of the socially constructed nature of gender inequalities. However, because migration is mainly an act of survival and social mobility, migrants do not direct their aims towards achieving greater gender equality.

Using intersectionality

Intersectionality has provided a useful framework for understanding migrants' realities. It has been useful in terms of better conceptualising the ways in which migrants' experiences in Buenos Aires were not only gendered, but also racialised and informed by the ethnic background of the people I was talking to. Intersectionality allowed me to move beyond a focus on gender and understand how migrants' experiences were shaped by the xenophobia and racism they encountered in Buenos Aires, particularly during the 1990s.

One of the main critiques of intersectionality relates to its lack of an explicit methodology (McCall 2005). Methodologically, I have taken the approach further, by going beyond the life story. While life stories informed the basis of this research, they were complemented by two community surveys and non-participant observations that allowed me to scale up some of the findings. This does not mean that the findings are generalisable to all migrants, not even within the Latin American region. However, the mixed-method approach has given me better traction at the meso-level of analysis, in terms of better understanding how social networks have evolved and the gendered access to labour markets, which a focus on individual life stories would not have provided. Clearly, this is not a 'how to' book and there is still much to say and think about in terms of the methodology of intersectionality. However, this research has gone a step further in exploring how we could use intersectionality at the meso-level of analysis.

This study also went beyond the usual ways in which others have used intersectionality in migration studies, in terms of who has been part of the study: I have included men. This was necessary to be able to talk about gender relations and not focus on women only, although the analysis does pay more attention to women's autonomy and empowerment. This is because of its original framing in feminist theory and a recognition that gender relations are usually heavily skewed towards providing men with more privileges and power.

Intersectionality has been invaluable in terms of analysing how women trade gender gains for upward social mobility. Without an understanding of class positions and the ways in which migrants were already discriminated against as ex-miners when they first moved to Cochabamba, it would have been impossible to understand women's strategies and negotiations of the gains they made while abroad. However, this has also been possible because of the longitudinal, historically-informed study that has paid attention to translocal cultural contexts and the forms that gender, class, and ethnicity take in the study sites.

I have worried about colonising the reality of the people that took their time to share their experiences with me, not least because I am using a framework that originated elsewhere. There are no simple answers to this conundrum. I have always been open about the feminist objectives and interests of this research. Those who did not share these objectives were obviously free not to participate and some, indeed, refused to participate. However, I also feel that the use of intersectionality has allowed me to get closer to the migrants' experiences by also paying attention to the ways in which their stories about gender are also shaped by class and ethnicity.

As we move towards a better understanding of regional migrations and the complex relationship between regional and global migrations, migration studies that adopt an approach similar to the one in this book will become more common. One hopes that researchers will continue to treasure lesser-known migration experiences, one that may not attract the attention of policy-makers sitting in Paris, Rome, or Washington, DC. I hope I did some justice, at least to make the story of this community and the experiences of these women and men migrants more accessible to others. My hope is, therefore, that through this book, I remained true to both the people who talked to me, but also to the framework's feminist origins and the ideas that have informed this research.

References

Abadan-Unat, N. 1977. Implications of migration on emancipation and pseudo-emancipation of Turkish women. *The International Migration Review* 11(1): 31–57.

Abdulrahim, D. 1993. Defining gender in a second exile: Palestinian women in West Berlin. In *Migrant Women: Crossing Boundaries and Changing Identities*, ed. G. Buijs, 55–82. Oxford: Berg.

Abella, M. 1995. Sex selectivity of migration regulations governing international migration in southern and south-eastern Asia. In *International Migration Policies and the Status of Female Migrants*. New York: UN.

Afshar, H. and M. Maynard. 1994. *The Dynamics of 'Race' and Gender: Some Feminist Interventions*. London: Taylor & Francis.

Ahmed, S. 2004. *The Cultural Politics of Emotion*. Edinburgh: Edinburgh University Press.

Allen, C.J. 2002. *The Hold Life Has: Coca and Cultural Identity in an Andean Community*. Washington, DC; London: Smithsonian Institution Press.

Altintas, E. and O. Sullivan. 2016. Fifty years of change updated: Cross-national gender convergence in housework. *Demographic Research* 35: 455–470.

Anderson, B. 1983. *Imagined Communities: Reflections on the Origin and Spread of Nationalism*. London: Verso.

Anderson, B. 2000. *Doing the Dirty Work?: the Global Politics of Domestic Labour*. New York: Zed Books.

Anthias, F. 1983. Sexual divisions and ethnic adaptation: The case of Greek-Cypriot women. In *One Way Ticket: Migration and Female Labour*, ed. A. Phizacklea. London: Routledge and Kegan Paul.

Anthias, F. 2002. Beyond feminism and multiculturalism: Locating difference and the politics of location. *Women's Studies International Forum* 25(3): 275–286.

Anthias, F. and N. Yuval-Davis. 1983. Contextualizing feminism: Gender, ethnic and class divisions. *Feminist Review* 15(November): 62–75.

Arruzza, C., T. Bhattacharya, and N. Fraser. 2018. Notes for a feminist manifesto. *New Left Review* 114: 113–134.

Baby-Collin, V. 2014. El servicio doméstico en trayectorias de bolivianas migrantes a España. In *Trabajadoras en la sombra. Dimensiones del servicio doméstico latinoamericano*, eds. S. Durin, M.E. de la O, and S. Bastos, 485–508. Mexico: Publicaciones de la casa chata CIESAS.

Baby-Collin, V. and G. Cortes. 2014. New trends of the Bolivian migration in a context of crisis. *Revista CIDOB d'Afers Internacionals* 106–107: 61–84.

Baby-Collin, V., G. Cortes, and S. Sassone. 2008. Mujer, movilidad y territorialización: Análisis cruzado de las migraciones internacionales en México y Bolivia. In *Migración transnacional de los Andes a Europa y Estados Unidos*, eds. H. Godard and G. Sandoval, 135–166. Lima; La Paz: IFEA/PIEB/IRD.

Bailey, A.J. 2013. Migration, recession and an emerging transnational biopolitics across Europe. *Geoforum* 44: 202–210.

Balán, J. 1995. Household economy and gender in international migration: the case of Bolivians in Argentina.In *International Migration Policies and the Status of Female Migrants: Proceedings of the United Nations Expert Group Meeting on International Migration Policies and the Status of Female Migrants*, ed. UN. San Miniato, Italy: UN.

Baldassar, L., C.V. Baldock, and R. Wilding. 2007. *Families Caring Across Borders: Migration, Ageing, and Transnational Caregiving*. Basingstoke; New York: Palgrave Macmillan.

Barnes, J.A. 1954. Class and committees in a Norwegian island parish. *Human Relations* 7: 39–58.

Barrios de Chungara, D., with M. Viezzer, and V. Ortiz. 1979. *Let me speak! Testimony of Domitila, a woman of the Bolivian mines*. London: Monthly Review Press.

Basch, L.G., N. Glick Schiller, and C. Szanton Blanc. 1994. *Nations Unbound: Transnational Projects, Postcolonial Predicaments, and Deterritorialized Nation-states*. London and New York: Routledge.

Bastia, T. 2007. From mining to garment workshops: Bolivian migrants in Buenos Aires. *Journal of Ethnic and Migration Studies* 33(4): 655–669.

Bastia, T. 2009. Women's migration and the crisis of care: Grandmothers caring for grandchildren in urban Bolivia. *Gender & Development* 17(3): 389–401.

Bastia, T. 2011a. Migration as protest? Negotiating gender, class, and ethnicity in urban Bolivia. *Environment and Planning A* 43(7): 1514–1529.

Bastia, T. 2011b. Should I stay or should I go? Return migration in times of crises. *Journal of International Development* 23(4): 583–595.

Bastia, T. 2012. 'I am going, with or without you': Autonomy in Bolivian transnational migrations. *Gender, Place & Culture* 20(2): 160–177.

Bastia, T. 2013a. *Migration and Inequality*. New York; Abingdon: Routledge.

Bastia, T. 2013b. Migration and inequality: An introduction. In *Migration and Inequality*, ed. T. Bastia. New York; Abingdon: Routledge.

Bastia, T. 2014. Intersectionality, migration and development. *Progress in Development Studies* 14(3): 237–248.

Bastia, T. 2015a. 'Looking after granny': A transnational ethic of care and responsibility. *Geoforum* 64: 121–129.

Bastia, T. 2015b. Transnational migration and urban informality: Ethnicity in Buenos Aires' informal settlements. *Urban Studies* 52(10): 1810–1825.

Bastia, T. 2018. Transnational migration and the gendered right to the city in Buenos Aires. *Cities* 76: 18–22.

Bastia, T. and S. McGrath. 2011. Temporality, migration and unfree labour: Migrant garment workers. In *Manchester Papers in Political Economy*. Manchester: University of Manchester.

Bastia, T. and J. Montero Bressán. 2018. Between a guest and an okupa: Migration and the making of insurgent citizenship in Buenos Aires' informal settlements. *Environment and Planning A: Economy and Space* 50(1): 31–50.

Bastia, T. and M. vom Hau. 2014. Migration, race and nationhood in Argentina. *Journal of Ethnic and Migration Studies* 40(3): 475–492.

Benencia, R. 1997. De peones a patrones quinteros: movilidad social de familias bolivianas en la periferia bonaerense. *Estudios Migratorios Latinoamericanos* 12(35): 63–101.

Benencia, R. and A. Gazzotti. 1995. Migración limítrofe y empleo: precisiones e interrogantes. *Estudios Migratorios Latinoamericanos* 10(31): 573–611.

Benencia, R. and G. Karasik. 1995. *Inmigración Limítrofe: Los Bolivianos en Buenos Aires*. Buenos Aires: Centro Editor de América Latina.

Benería, L. and G. Sen. 1982. Class and gender inequalities and women's role in economic development: Theoretical and practical implications. *Feminist Studies* 8(1): 157–176.

Bianchi, S.M., L.C. Sayer, M.A. Milkie, and J.P. Robinson. 2012. Housework: Who did, does or will do it, and how much does it matter? *Social Forces* 91(1): 55–63.

Bilecen, B., M. Gamper, and M.J. Lubbers. 2018. The missing link: Social network analysis in migration and transnationalism. *Social Networks* 53: 1–3.

Black, R., C. Natali, and J. Skinner. 2005. Migration and inequality. In *World Development Report: Background Papers*. Washington, DC: World Bank.

Bloch, A. 2017. *Sex, Love, and Migration: Postsocialism, Modernity, and Intimacy from Istanbul to the Arctic*. New York: Cornell University Press.

Bocangel Jerez, D. 2001. Bolivia: Estudio regional/nacional sobre pequeña minería y artesanal. In *Proyecto MMSD*. London: IIED andWBCSD.

Boccagni, P. 2013. Migration and the family transformations it "leaves behind": A critical view from Ecuador. *The Latin Americanist* 57(4): 3–24.

Boehm, D.A. 2011. Deseos y dolores: Mapping desire, suffering, and (dis)loyalty within transnational partnerships. *International Migration* 49(6): 95–106.

Boehm, D.A. 2012. *Intimate Migrations: Gender, Family, and Illegality among Transnational Mexicans*. New York: New York University Press.

Boesten, J. 2010. *Intersecting Inequalities: Women and Social Policy in Peru, 1990–2000*. University Park, PA: Pennsylvania State University Press.

Bondi, L. 2004. 10th Anniversary Address: For a feminist geography of ambivalence. *Gender, Place & Culture: A Journal of Feminist Geography* 11(1): 3–15.

Bonifacio, G.T. 2012. *Feminism and Migration: Cross-cultural Engagements*. Dordrecht; London: Springer.

Bourdieu, P. 1977. *Outline of a Theory of Practice*. Cambridge: Cambridge University Press.

Bourque, S.C. and K.B. Warren. 1981. *Women of the Andes: Patriarchy and Social Change in Two Peruvian Towns*. Ann Arbor, MI: University of Michigan Press.

Boyle, P.J. and K. Halfacree. 1999. *Migration and Gender in the Developed World*. London; New York: Routledge.

Brah, A. and A. Phoenix. 2004. Ain't I a woman? Revisiting intersectionality. *Journal of Women's International Studies* 5(3): 75–86.

Braidotti, R. 1994. *Nomadic Subjects: Embodiment and Sexual Difference in Contemporary Feminist Theory*. New York: Columbia University Press.

Brettell, C.B. and R.J. Simon. 1986. Immigrant women: An introduction. In *International Migration: The Female Experience*, eds. R.J. Simon and C.B. Brettell. Totowa, NJ: Rowman and Allanheld.

Buijs, G. 1993. *Migrant Women: Crossing Boundaries and Changing Identities*. Oxford: Berg.

Buitelaar, M. 2006. 'I am the ultimate challenge'. *European Journal of Women's Studies* 13(3): 259–276.

Burawoy, M. 2000. *Global Ethnography: Forces, Connections, and Imaginations in a Postmodern World*. Berkeley, CA: University of California Press.

Burman, E. 2003. From difference to intersectionality: Challenges and resources. *European Journal of Psychotherapy & Counselling* 6(4): 293–308.

Butler, J. 1990. *Gender Trouble: Feminism and the Subversion of Identity*. New York; London: Routledge.

Calderon, F. and A. Rivera. 1984. *La Cancha: Una gran feria campesina en la ciudad de Cochabamba*. Cochabamba: CERES Centro de Estudios de la Realidad Económica y Social.

Canessa, A. 2007. Who is indigenous? Self-identification, indigeneity, and claims to justice in contemporary Bolivia. *Urban Anthropology and Studies of Cultural Systems and World Economic Development* 36(3): 195–237.

Canessa, A. 2012. *Intimate Indigeneities: Race, Sex, and History in the Small Spaces of Andean life*. Durham, NC: Duke University Press.

Çaro, E., A. Bailey and L.J.G. Van Wissen. 2012. Negotiating between patriarchy and emancipation: Rural-to-urban migrant women in Albania. *Gender, Place and Culture* 19(4): 479–493.

Castles, S. 2010. Understanding global migration: A social transformation perspective. *Journal of Ethnic and Migration Studies* 36(10): 1565–1586.

Cerrutti, M. 2009. Gender and intra-regional migration in South America. In *Human Development Reports Research Paper 2009/12*: UNDP.

Chaney, E.M., M.G. Castro, and M.L. Smith. 1989. *Muchachas No More: Household Workers in Latin America and the Caribbean*. Philadelphia, PA: Temple University Press.

Chant, S. 1992. *Gender and Migration in Developing Countries*. London: Belhaven.

Chant, S. 2009. The 'feminisation of poverty' in Costa Rica: To what extent a conundrum? *Bulletin of Latin American Research* 28(1): 19–43.

Chant, S. and S. Radcliffe. 1992. Migration and development: The importance of gender. In *Gender and Migration in Developing Countries*, ed. S. Chant. London: Belhaven Press.

Chow, E.N.-L. 1996. Introduction: Transforming knowledge: Race, class, gender. In *Race, Class and Gender: Common Bonds, Different Voices*, eds. E.N.-L. Chow, D. Wilkinson, and M. Baca Zinn. London: Sage.

Chow, Yiu F. 2011. Moving, sensing intersectionality: A case study of Miss China Europe. *Signs* 36(2): 411–436.

Collins, P.H. 1990. *Black Feminist Thought: Knowledge, Consciousness, and the Politics of Empowerment*. Boston, MA; London: Unwin Hyman.

Constable, N. 1997. Sexuality and discipline among Filipina domestic workers in Hong Kong. *American Ethnologist* 24(3): 539–558.

Constable, N. 2003. *Romance on a Global Stage: Pen Pals, Virtual Ethnography, and 'Mail-order' Marriages*. Berkeley, CA; London: University of California Press.

Constable, N. 2007. *Maid to Order in Hong Kong: Stories of Migrant Workers*. Ithaca, NY: Cornell University Press.

Constable, N. 2014. *Born Out of Place: Migrant Mothers and the Politics of International Labor*. Berkeley, CA: University of California Press.

Correa, V. and M.I. Pacecca. 1999. Las mujeres migrantes en la ciudad de Buenos Aires: Características socio-laborales. In *Sub-proyecto: Aspectos Laborales de los Migrantes en América Latina*. Buenos Aires: IOM.

Cortes, G. 2004. *Partir para Quedarse: Supervivencia y Cambio en las Sociedades Campesinas Andinas.* La Paz: Plural.

Cortes, G. 2011. The making of the transnational family: A diachronic approach of migration platforms and dispersion of rural Bolivian families. *Autrepart* 57–58(1): 95–110.

Cortés Cartellanos, P. 2005. *Mujeres Migrantes de América Latina y el Caribe: Derechos Humanos, Mitos y Duras Realidades.* Santiago de Chile: CEPAL.

Courtis, C. 2000. *Construcciones de Alteridad: Discursos Cotidianos sobre la Inmigración Coreana en Buenos Aires.* Buenos Aires: EUDEBA.

Courtis, C. and M.I. Pacecca. 2010. Género y trayectoria migratoria: Mujeres migrantes y trabajo doméstico en el Área Metropolitana de Buenos Aires. *Papeles de Población* 16: 155–185.

Crenshaw, K. 1991. Mapping the margins: Intersectionality, identity politics, and violence against women of color. *Stanford Law Review* 43(6): 1241–1299.

Crivello, G. 2011. 'Becoming somebody': Youth transitions through education and migration in Peru. *Journal of Youth Studies* 14(4): 395–411.

Curran, S.R., S. Shafer, K.M. Donato, and F. Garip. 2006. Mapping gender and migration in sociological scholarship: Is it segregation or integration? *International Migration Review* 40(1): 199–223.

Dandler, J. and C. Medeiros. 1988. Temporary migration from Cochabamba, Bolivia to Argentina: Patterns and impact in sending areas. In *When Borders Don't Divide: Labour Migration and Refugee Movements in the Americas*, ed. P.R. Pessar, 8–41. New York: Centre for Migration Studies.

Datta, K., C. McIlwaine, Y. Evans, J. Herbert, J. May and J. Wills. 2010. A migrant ethic of care: Negotiating care and caring among migrant workers in London's low-pay economy. *Feminist Review* 94(1): 93–116.

Davis, A.Y. 1981. *Women, Race, & Class.* New York: Random House.

Daya, S. 2009. Embodying modernity: Reading narratives of Indian women's sexual autonomy and violation. *Gender, Place & Culture: A Journal of Feminist Geography* 16(1): 97–110.

de Beauvoir, S. 1993. *The Second Sex.* London: David Campbell.

de Haas, H. 2012. The migration and development pendulum: A critical view on research and policy. *International Migration* 50(3): 8–25.

de Haas, H. and A. van Rooij. 2010. Migration as emancipation? The impact of internal and international migration on the position of women left behind in rural Morocco. *Oxford Development Studies* 38(1): 43–62.

de Jong, G.F. 2000. Expectations, gender, and norms in migration decision-making. *Population Studies: A Journal of Demography* 54(3): 307–319.

de la Torre Avila, L. 2006. *No Llores, Prenda, Pronto Volvere. Migracion, Movilidad Social, Herida Familiar y Desarrollo.* La Paz: PIEB.

de la Torre Avila, L. and Y. Alfaro Aramayo. 2007. *La Cheqanchada: Caminos y Sendas de Desarrollo en los Municipios Migrantes de Arbieto y Toco.* La Paz: PIEB.

Deere, C.D. 1977. Changing social relations of production and Peruvian peasant women's work. *Latin American Perspectives* 4(1/2): 48–69.

di Leonardo, M. 1987. The female world of cards and holidays: Women, families, and the work of kinship. *Signs* 12(3): 440–453.

Donato, K.M. and D. Gabaccia. 2015. *Gender and International Migration: From the Slavery Era to the Global Age.* New York: Russell Sage Foundation.

Donato, K.M., D. Gabaccia, J. Holdaway, M. Manalansan and P.R. Pessar. 2006. A glass half full? Gender in migration studies. *International Migration Review* 40(1): 3–26.

Durand, J. and D.S. Massey. 2006. *Crossing the Border: Research from the Mexican Migration Project*. New York: Russell Sage Foundation.

Eastmond, M. 1993. Reconstructing life: Chilean refugee women and the dilemmas of exile. In *Migrant Women: Crossing Boundaries and Changing Identities*, ed. G. Buijs, 35–54. Oxford: Berg.

El Mundo. 2015. *El Mapa de la Violencia de Género en España*: El Mundo.

England, K. and I. Dyck. 2012. Migrant workers in home care: Routes, responsibilities, and respect. *Annals of the Association of American Geographers* 102(5): 1076–1083.

Escandell, X. and M. Tapias. 2009. Transnational lives, travelling emotions and idioms of distress among Bolivian migrants in Spain. *Journal of Ethnic and Migration Studies* 36(3): 407–423.

EU. 2009. *Clandestino Project: Final Report*, ed. E. Commission. Athens.

Faier, L. 2009. *Intimate Encounters: Filipina Women and the Remaking of Rural Japan*. Berkeley, CA; London: University of California Press.

Faier, L. 2011. Theorizing the intimacies of migration: Commentary on the emotional formations of transnational worlds. *International Migration* 49(6): 107–112.

Faist, T. 2000. *The Volume and Dynamics of International Migration and Transnational Social Spaces*. Oxford: Clarendon.

Ferrufino, C. 2007. *Los Costos Humanos de la Emigración*. La Paz: PIEB andCESU.

Fine, B. 1999. The developmental state is dead—Long live social capital? *Development and Change* 30(1): 1–19.

Fine, B. 2001. *Social Capital Versus Social Theory: Political Economy and Social Science at the Turn of the Millennium*. London: Routledge.

Fraser, N. 2005. Reframing justice in a globalising world. *New Left Review* 36: 69–88.

Fraser, N. 2007. Feminist politics in the age of recognition: A two-dimensional approach to gender justice. *Studies in Social Justice* 1(1): 23–35.

Gamburd, M.R. 2000. *The Kitchen Spoon's Handle: Transnationalism and Sri Lanka's Migrant Housemaids*. Ithaca, NY: Cornell University Press.

Gibson, K., L. Law and D. McKay. 2001. Beyond heroes and victims: Filipina contract migrants, economic activism and class transformations. *International Feminist Journal of Politics* 3(3): 365–386.

Gil Araujo, S. and T. González-Fernández. 2014. International migration, public policies and domestic work: Latin American migrant women in the Spanish domestic work sector. *Women's Studies International Forum* 46: 13–23.

Gill, L. 1997. Relocating class: Ex-miners and neoliberalism in Bolivia. *Critique of Anthropology* 17(3): 293–312.

Glick Schiller, N. 2011. A global perspective on migration and development. In *The Migration-Development Nexus: A Transnational Perspective*, eds. T. Faist, M. Fauser and P. Kivisto, 29–56. Basingstoke: Palgrave Macmillan.

Gogna, M. 1993. Empleadas domésticas en Buenos Aires. In *Muchacha, Cachifa, Criada, Empleada, Empregadinha, Sirvienta y ... Más Nada: Trabajadoras Domésticas en América Latina y el Caribe*, eds. E.M. Chaney and M.G. Castro. Buenos Aires: Editorial Nueva Sociedad.

Goldstein, D.M. 2005. Flexible justice. *Critique of Anthropology* 25(4): 389–411.

González, M. and S. Sassone. 2016. Mujeres migrantes, trabajo y empoderamiento: Bolivianas en una ciudad de la periferia globalizada. *Amérique Latine Histoire et Mémoire. Les Cahiers ALHIM* 31. https://journals.openedition.org/alhim/5453

González-Fernández, T. 2016. Entre nodos y nudos: Ambivalencias emocionales en la migración transnacional. Una aproximación etnográfica a las emociones a partir de familias transnacionales entre Bolivia y España. *Odisea. Revista de Estudios Migratorios* 3(5): 99–123.

Gray-Molina, G., W. Jiménez, E.P. de Rada and E. Yáñez. 1999. Pobreza y activos en Bolivia ¿Qué papel desempeña el capital social? *El Trimestre Económico* 66(3): 365–417.

Green, M. and V. Lawson. 2011. Recentring care: Interrogating the commodification of care. *Social & Cultural Geography* 12(6): 639–654.

Gregorio Gil, C. 1998. *Migración Femenina: Su Impacto en las Relaciones de Género.* Madrid: Narcea, S.A. de Ediciones.

Gregson, N., U. Kothari, J. Cream, C. Dwyer, S. Holloway, A. Maddrell and G. Rose. 1997. Gender in feminist geography. In *Feminist Geographies: Explorations in Diversity and Difference*, ed. W.a.G.S. Group. Harlow: Longman.

Grimson, A. 1999. *Relatos de la Diferencia y la Igualdad: Los Bolivianos en Buenos Aires.* Buenos Aires, Argentina: EUDEBA: Federación Latinoamericana de Facultades de Comunicación Social.

Grimson, A.and E. Paz Soldan. 2000. *Migrantes Bolivianos en la Argentina y los Estados Unidos.* La Paz: UNDP.

Gugler, J. 1997. Gender and rural-urban migration: Regional contrasts and the gender transition. In *Migration and Gender: Place, Time and People Specific*, eds. Joan Fairhurst et al. Pretoria: IGU Commission on gender and geography.

Guijt, I. and M.K. Shah. 1998. *The Myth of Community: Gender Issues in Participatory Development.* London: Intermediate Technology Publications.

Hagan, J.M. 1998. Social networks, gender, and immigrant incorporation: Resources and constraints. *American Sociological Review* 63: 55–67.

Harris, O. 1994. Condor and the bull: The ambiguities of masculinity in Northern Potosi. In *Sex and Violence: Issues in Representation and Experience*, eds. P. Harvey and P.G. Gow. London: Routledge.

Harvey, P. 1994. Domestic violence in the Peruvian Andes. In *Sex and Violence: Issues in Representation and Experience*, eds. P. Harvey and P.G. Gow. London: Routledge.

Herrera, G. 2013. *Lejos de Tus Pupilas: Familias Transnacionales, Cuidados y Desigualdad Social en Ecuador.* Quito: Onu-Muejres, FLACSO.

Herrera, G. forthcoming. Care and migration. In *Routledge Handbook of Migration and Development*, eds. T. Bastia and R. Skeldon. London: Routledge.

Hill Collins, P. 2009. *Black Feminist Thought: Knowledge, Consciousness, and the Politics of Empowerment.* New York: Routledge.

Hinojosa, A. 2008a. España en el itinerario de Bolivia. Migración transnational, género y familia en Cochabamba. In *Las Migraciones en America Latina*, ed. S. Novick, 93–112. Buenos Aires: CLACSO.

Hinojosa, A. 2008b. Transnacionalismo y multipolaridad en los flujos migratorios de Bolivia. Familia, comunidad y nación en dinámicas globales. In *Migración Transnacional de los Andes a Europa y Estados Unidos*, eds. H.R. Godard and G. Sandoval, 77–101. Lima: Institut français d'études andines. IFEA – Institut de recherche pour le développement. IRD – Programa de investigación estratégica en Bolivia. PIEB.

Hinojosa, A. 2009a. *Buscando la Vida: Familias Bolivianas Transnacionales en España*. La Paz: CLACSO; Fundación PIEB.

Hinojosa, A. 2009b. *Migracion Transnacional y Sus Efectos en Bolivia*. La Paz: PIEB, Programa de Investigacion Estrategica en Bolivia.

Hinojosa, A. forthcoming. Migration corridor Bolivia-Argentina. In *Routledge Handbook of Migration and Development*, eds. T. Bastia and R. Skeldon. London: Routledge.

Hirschmann, N.J. 2003. *The Subject of Liberty: Toward a Feminist Theory of Freedom*. Princeton, NJ; Oxford: Princeton University Press.

Hochschild, A.R. 2000. Global care chains and emotional surplus value. In *On the Edge: Living with Global Capitalism*, eds. A. Giddens and W. Hutton. London: Jonathan Cape.

Hochschild, A.R. and A. Machung. 1989. *The Second Shift: Working Parents and the Revolution at Home*. London: Piatkus.

Hondagneu-Sotelo, P. 1994. Overcoming Patriarchal Constraints: The Reconstruction of Gender Relations among Mexican Immigrant Women and Men. *Gender & Society* 6(3): 393–415.

Hondagneu-Sotelo, P. 1994. *Gendered Transitions: Mexican Experiences of Immigration*. Berkeley, CA; London: University of California Press.

Hondagneu-Sotelo, P. 2000. Feminism and migration. *Annals of the American Academy of Political and Social Science* 571: 107–120.

Hondagneu-Sotelo, P. and E. Avila. 1997. "I'm here, but I'm there": The meanings of Latina transnational motherhood. *Gender Society* 11(5): 548–571.

hooks, b. 1997. Feminism: A movement to end sexist oppression.In *Feminisms*, eds. A. Kemp and J. Squires. Oxford: Oxford University Press.

hooks, b. 1999. *Feminist Theory: From Margin to Center*. Cambridge, MA: South End Press.

Horne, R.M., M.D. Johnson, N.L. Galambos and H.J. Krahn. 2018. Time, money, or gender? Predictors of the division of household labour across life stages. *Sex Roles* 78(11): 731–743.

Hugo, G. 2002. Effects of international migration on the family in Indonesia. *Asian and Pacific Migration Journal* 11(1): 13–46.

ILO. 2014. *New Law Leads to New Life for Migrant Domestic Workers*.

INDEC. 1994. Estimaciones y proyecciones de población 1950–2050: Total del país. In *Estudios INDEC, no.23*. Buenos Aires: INDEC.

INDEC. 1997. La migración internacional en la Argentina: Sus características e impacto. In *Estudios INDEC, no.29*. Buenos Aires: INDEC.

INE. 2010. *Padrón de población, Instituto Nacional de Estadística, Spain*, www.ine.es

INE. 2017. *Cifras de Poblacion a 1 de enero 2017. Estadística de migraciones 2016. Datos provisionales*. Madrid: Instituto Nacional de Estadística.

INE, Instituto Nacional de Estadística. 2015. *Bolivia: Feminicidio por departamento*. La Paz: Estado Plurinacional de Bolivia.

INE, Instituto Nacional de Estadística. 2016. *Violencia doméstica*. La Paz: Estado Plurinacional de Bolivia.

Ioe, C. 1991. *Foreign Women in Domestic Service in Madrid, Spain*, ed. W.E. Programme. Geneva: International Labour Organisation.

Isbell, B.J. 1979. *To Defend Ourselves: Ecology and Ritual in an Andean Village*. Austin: University of Texas.

Jackson, C. 2003. Gender analysis of land: Beyond land rights for women? *Journal of Agrarian Change* 3(4): 453–480.

Janjuha-Jivraj, S. 2003. The sustainability of social capital within ethnic networks. *Journal of Business Ethics* 47(1): 31–43.

Jones, R.C. and L. de la Torre. 2011. Diminished tradition of return? Transnational migration in Bolivia's Valle Alto. *Global Networks* 11(2): 180–202.

Kabeer, N. 1999. *The Conditions and Consequences of Choice: Reflections on the Measurement of Women's Empowerment*. Geneva: UNRISD.

Kabeer, N. 2000. *The Power to Choose: Bangladeshi Women and Labour Market Decisions in London and Dhaka*. London: Verso.

Kofman, E. 1999. Female 'birds of passage' a decade later: Gender and immigration in the European Union. *International Migration Review* 33(2): 269–299.

Kofman, E. 2012. Rethinking care through social reproduction: Articulating circuits of migration. *Social Politics* 19(1): 142–162.

Kofman, E. and P. Raghuram. 2012. Women, migration, and care: Explorations of diversity and dynamism in the global south. *Social Politics: International Studies in Gender, State & Society* 19(3): 408–432.

Kofman, E. and P. Raghuram. 2015. *Gendered Migrations and Global Social Reproduction*. London: Palgrave.

Kohl, B. 2006. Challenges to neoliberal hegemony in Bolivia. *Antipode* 38(2): 304–326.

Kornienko, O., V. Agadjanian, C. Menjívar and N. Zotova. 2018. Financial and emotional support in close personal ties among Central Asian migrant women in Russia. *Social Networks* 53: 125–135.

Kosnik, K. 2011. Sexuality and migration studies: The invisible, the oxymoronic and heteronormative othering. In *Framing Intersectionality: Debates on a Multi-faceted Concept in Gender Studies*, eds. H. Lutz, M.T. Herrera Vivar and L. Supik. Farnham: Ashgate.

Kothari, U. 2002. *Migration and Chronic Poverty*. Chronic Poverty Research Centre.

Kothari, U. 2003. Staying put and staying poor? *Journal of International Development* 15: 645–657.

Kothari, U. 2008. Global peddlers and local networks: Migrant cosmopolitanisms. *Environment and Planning D: Society and Space* 26(3): 500–516.

Kyle, D. 1999. The Otavalo trade diaspora: Social capital and transnational entrepreneurship. *Ethnic and Racial Studies* 22(2): 422–446.

Kynsilehto, A. 2011. Negotiating intersectionality in highly educated migrant Maghrebi women's life stories. *Environment and Planning A* 43(7): 1547–1561.

Laurie, N. 1999. The shifting geographies of femininity and emergency work in Peru. In *Geographies of New Femininities*, eds. N. Laurie, C. Dwyer, S.L. Holloway and F. M. SmithHarlow: Longman.

Lawson, V. 2007. Geographies of care and responsibility. *Annals of the Association of American Geographers* 97(1): 1–11.

Lind, A. 2010. *Development, Sexual Rights and Global Governance*. London: Routledge.

Lorde, A. 1985. *I Am Your Sister: Black Women Organizing Across Sexualities*. New York: Kitchen Table, Women of Color Press.

Ludvig, A. 2006. Differences between women? Intersecting voices in a female narrative. *European Journal of Women's Studies* 13(3): 245–258.

Lutz, H. 2010. Gender in the migratory process. *Journal of Ethnic and Migration Studies* 36(10): 1647–1663.

Lutz, H., M.T. Herrera Vivar and L. Supik. 2011. *Framing Intersectionality: Debates on a Multi-faceted Concept in Gender Studies*. Farnham: Ashgate.

Mackenzie, C. and N. Stoljar. 2000. *Relational Autonomy: Feminist Perspectives on Autonomy, Agency, and the Social Self.* New York; Oxford: Oxford University Press.

Magliano, M.J. 2007. Migración de mujeres bolivianas hacia Argentina: Cambios y continuidades en las relaciones de género. *Amérique Latine Histoire et Mémoire. Les Cahiers ALHIM* 14. https://journals.openedition.org/alhim/2102

Mahler, S.J. and P.R. Pessar. 2006. Gender matters: Ethnographers bring gender from the periphery toward the core of migration studies. *International Migration Review* 40(1): 27–63.

Martínez Buján, R. 2010. *Bienestar y Cuidados: El Oficio del Cariño: Mujeres Inmigrantes y Mayores Nativos.* Madrid: Consejo Superior de Investigaciones Científicas.

Massey, D.S. 1999. International migration at the dawn of the twenty-first century: The role of the state. *Population and Development Review* 25(2): 303–322.

Massey, D.S., J. Arango, G. Hugo, A. Kouaouci, A. Pellegrino and J.E. Taylor. 1993. Theories of international migration: A review and appraisal. *Population and Development Review* 19(3): 431–466.

Matsuda, M.J. 1991. Beside my sister, facing the enemy: Legal theory out of coalition. *Stanford Law Review* 43(6): 1183–1192.

Maynard, M. 1994. "Race", gender and the concept of "difference" in feminist thought. In *The Dynamics of 'Race' and Gender*, eds. H. Afshar and M. Maynard. London: Taylor and Francis.

McCall, L. 2005. The complexity of intersectionality. *Signs: Journal of Women in Culture and Society* 5(3): 29.

McDowell, L. 2008. Thinking through class and gender in the context of working class studies. *Antipode* 40(1): 20–24.

McEwan, C. 2001. Postcolonialism, feminism and development: Intersections and dilemmas. *Progress in Development Studies* 1(2): 93–111.

McIlwaine, C. 2010. Migrant machismos: Exploring gender ideologies and practices among Latin American migrants in London from a multi-scalar perspective. *Gender, Place & Culture* 17(3): 281–300.

McIlwaine, C. and A. Bermudez. 2011. The gendering of political and civic participation among Colombian migrants in London. *Environment and Planning A* 43(7): 1499–1513.

McIlwaine, C. and M. Ryburn. 2018. Diversities of international and transnational migration in and beyond Latin America. In *The Routledge Handbook of Latin American Development*, eds. J. Cupples, M. Palomino-Schalscha and M. Prieto. Abingdon: Routledge.

Menjívar, C. 2000. *Fragmented Ties: Salvadoran Immigrant Networks in America.* Berkeley, CA: University of California Press.

Menjívar, C. 2005. Migration and refugees. In *A Companion to Gender Studies*, eds. P. Essed, D.T. Goldberg and A. Kobayashi. Malden, MA; Oxford: Blackwell.

Mills, M.B. 1997. Contesting the margins of modernity: Women, migration, and consumption in Thailand. *American Ethnologist* 24(1): 37–61.

Ministerio del Interior and DNM. 2010. *Patria Grande: Programa Nacional de Normalizacion Documentaria Migratoria. Informe Estadistico.* Buenos Aires: Ministerio del Interior, Direccion Nacional de Migraciones.

Mohanty, C.T. 1988. Under western eyes: Feminist scholarship and colonial discourses. *Feminist Review* 30: 61–88.

Molyneux, M. 2007. Change and continuity in social protection in Latin America: Mothers at the service of the state? In *Gender and Development Programme*. Geneva: United Nations Research Institute for Social Development

Montero Bressán, J. 2011. Neoliberal fashion: The political economy of sweatshops in Europe and Latin America. In *Geography*. Durham: University of Durham.

Montero Bressán, J. and A. Arcos. 2016. How do migrants respond to labour abuses in local sweatshops? *Antipode 49*(2): 437–454.

Morgan, R. 1970. *Sisterhood is Powerful*. New York: Random House.

Morokvasic, M. 1984. Birds of passage are also women... *International Migration Review* 18(4): 886–907.

Mugarza, S. 1985. Presencia y ausencia boliviana en la ciudad de Buenos Aires. *Estudios Migratorios Latinoamericanos* 1(1): 98–106.

Näre, L. 2011. The moral economy of domestic and care labour: Migrant workers in Naples, Italy. *Sociology* 45(3): 396–412.

Nash, J. 1976. Mi vida en las minas: La autobiografía de una mujer boliviana. *Estudios Andinos* 5(1): 139–150.

Nash, J. 1993. *We Eat the Mines and the Mines Eat Us: Dependency and Exploitation in Bolivian Tin Mines*. New York; Guildford: Columbia University Press.

Nash, J. 2008. Re-thinking intersectionality. *Feminist Review* 89: 1–15.

Nelson, D.R. 2015. Migration and networks. In *Complexity and Geographical Economics: Topics and Tools*, eds. P. Commendatore, S. Kayam and I. Kubin, 141–164. Cham: Springer International Publishing.

Núñez del Prado Béjar, D. 1975a. El poder de decisión de la mujer quechua andina. *América Indígena* 35: 623–630.

Núñez del Prado Béjar, D. 1975b. El rol de la mujer campesina quechua. *América Indígena* 35: 391–401.

Orellana Halkyer, R. n.d. *Entre Lana y Cemento: Inmigrantes Bolivianos en Argentina*, available from CERES library, Bolivia.

Palloni, A., D.S. Massey, M. Ceballos, K. Espinosa and M. Spittel. 2001. Social capital and international migration: A test using information on family networks. *American Journal of Sociology* 106(5): 1262–1298.

Pappas-DeLuca, K. 1999. Transcending gendered boundaries: Migration for domestic labour in Chile.In *Gender, Migration and Domestic Service*, ed. J. Momsen. London: Routledge.

Parella, S. 2003. *Mujer, Inmigrante y Trabajadora: La Triple Discriminación*. Barcelona: Anthropos.

Parella, S. 2011. Familia transnacional y redefinición de los roles de género: El caso de la migración boliviana en España. *Papers, Revista de Sociologia,* 97(3): 661–684.

Parella, S., A. Petroff and O. Serradell Pumareda. 2014. Programas de retorno voluntario en Bolivia y España en contextos de crisis. *Revista CIDOB d'Afers Internacionals* 106/107: 171–192.

Parella, S. and S. Samper. 2007. Factores explicativos de los discursos y estrategias de conciliación del ámbito laboral y familiar de las mujeres inmigradas no comunitarias en España. *Papers, Revista de Sociologia* 85: 157–175.

Parreñas, R.S. 2005. *Children of Global Migration: Transnational Families and Gendered Woes*. Stanford, CA: Stanford University Press.

Parreñas, R.S. 2008. *The Force of Domesticity: Filipina Migrants and Globalization*. New York; London: New York University Press.

Paulson, S. 2010. Headless families and detoured men: Off the straight path of modern development in Bolivia. In *Development, Sexual Rights and Global Governance*, ed. A. Lind. London; New York: Routledge.

Pedraza, S. 1991. Women and migration: The social consequences of gender. *Annual Review of Sociology* 17: 303–325.

Pessar, P.R. 1999. Engendering migration studies: The case of new immigrants in the United States. *American Behavioral Scientist* 42(4): 577–600.

Pessar, P.R. 2005. Women, gender, and international migration across and beyond the Americas: Inequalities and limited empowerment. In *Expert Group Meeting on International Migration and Development in Latin America and the Caribbean*. Mexico City: UN, Population Division.

Pessar, P.R. and S.J. Mahler. 2003. Transnational migration: Bringing gender in. *International Migration Review* 37(3): 812–846.

Phillips, N. 2009. Migration as development strategy? The new political economy of dispossession and inequality in the Americas. *Review of International Political Economy* 16(2): 231–259.

Phizacklea, A.E. 1983. *One Way Ticket: Migration and Female Labour*. London: R.K. P.

Pieterse, J.N. 2003. Social capital and migration. *Ethnicities* 3(1): 29–58.

Piper, N. 2008. Feminisation of migration and the social dimensions of development: The Asian case. *Third World Quarterly*, 29(7): 1287–1303.

Piper, N. and M. Roces. 2003. *Wife or Worker? Asian Marriage and Migration*. Lanham, MD; Oxford: Rowman and Littlefield.

Portes, A. 2010. Migration and social change: Some conceptual reflections. *Journal of Ethnic and Migration Studies* 36(10): 1537–1563.

Portes, A., L.E. Guarnizo and P. Landolt. 1999. The study of transnationalism: Pitfalls and promise of an emergent research field. *Ethnic and Racial Studies* 22(2): 217–237.

Portes, A. and L. Jensen. 1989. The enclave and the entrants: Patterns of ethnic enterprise in Miami before and after Mariel. *American Sociological Review* 54(6): 929–949.

Portes, A. and P. Landolt. 2000. Social capital: Promise and pitfalls of its role in development. *Journal of Latin American Studies* 32(2): 529–547.

Pratt, G. 1999. From registered nurse to registered nanny: Discursive geographies of Filipina domestic workers in Vancouver, B.C. *Economic Geography* 75(3): 215–236.

Pratt, G. 2009. Circulating sadness: Witnessing Filipina mothers' stories of family separation. *Gender, Place & Culture* 16(1): 3–22.

Pratt, G. 2012. *Families Apart: Migrant Mothers and the Conflicts of Labor and Love*. Minneapolis, MN: University of Minnesota Press.

Pratt, G. and B. Yeoh. 2003. Transnational (counter) topographies. *Gender, Place & Culture* 10(2): 159–166.

Pratt, G. 2005. From migrant to immigrant: Domestic workers settle in Vancouver, Canada. In *A Companion to Feminist Geography*, ed. L. Nelson and J. Seager. Oxford: Blackwell.

Pribilsky, J. 2007. *La Chulla Vida: Gender, Migration, and the Family in Andean Ecuador and New York City*. Syracuse, NY: Syracuse University Press.

Prins, B. 2006. Narrative accounts of origins. *European Journal of Women's Studies* 13 (3): 277–290.

Radcliffe, S.A. 1993. The role of gender in peasant migration: Conceptual issues from the Peruvian Andes.In *Different Places, Different Voices: Gender and Development in Africa, Asia and Latin America*, ed. V. Kinnaird and. J. Momsen. London: Routledge.

Radcliffe, S.A. 2006. Development and geography: Gendered subjects in development processes and interventions. *Progress in Human Geography* 30(4): 524–532.

Radcliffe, S.A. 2015. *Dilemmas of Difference: Indigenous Women and the Limits of Postcolonial Development Policy*. Durham, NC: Duke University Press.

Rankin, K.N. 2002. Social capital, microfinance, and the politics of development. *Feminist Economics* 8(1): 1–24.

Ravenstein, E.G. 1885. The laws of migration. *Journal of the Royal Statistical Society* 48: 167–227.

Recchini de Lattes, Z. 1988. *Las Mujeres en Las Migraciones Internas e Internacionales, con Especial Referencia a América Latina*, ed. CENEP. Buenos Aires.

Regalsky, P. 2003. *Etnicidad y Clase: El Estado Boliviano y Las Estrategias Andinas de Manejo de Su Espacio*. La Paz: Plural.

Requena Gonzáles, S. 2017. Una mirada a la situación de la violencia contra la mujer en Bolivia. *Revista de Investigacion Psicologica* 17: 117–134.

Resurreccion, B.P. and H.T.V. Khanh. 2007. Able to come and go: Reproducing gender in female rural-urban migration in the Red River Delta. *Population, Space and Place* 13(3): 211–224.

Riaño, Y. 2011. Drawing new boundaries of participation: Experiences and strategies of economic citizenship among skilled migrant women in Switzerland. *Environment and Planning A* 43(7): 1530–1546.

Rigg, J. 2007. Moving lives: Migration and livelihoods in the Lao PDR. *Population, Space and Place* 13(3): 163–178.

Robinson, F. 1999. *Globalizing Care: Ethics, Feminist Theory, and International Relations*. Boulder, CO: Westview Press.

Rockefeller, S.A. 2010. *Starting from Quirpini: The Travels and Places of a Bolivian People*. Bloomington, IN: Indiana University Press.

Rogers, A. and S. Vertovec. 1995. *The Urban Context: Ethnicity, Social Networks and Situational Analysis*. Oxford: Berg.

Román, O. 2009. *Mientras No Estamos: Migraciòn de Mujeres-madres de Cochabamba a España*. Cochabamba: UMSS, CESU.

Rose, G. 1993. *Feminism and Geography: The Limits of Geographical Knowledge*. Cambridge: Polity.

Roseneil, S. and S. Budgeon. 2004. Cultures of intimacy and care beyond 'the family': Personal life and social change in the early 21st century. *Current Sociology* 52(2): 135–159.

Ryan, L. and A. D'Angelo. 2018. Changing times: Migrants' social network analysis and the challenges of longitudinal research. *Social Networks* 53: 148–158.

Ryburn, M. 2016. Living the Chilean dream? Bolivian migrants' incorporation in the space of economic citizenship. *Geoforum* 76: 48–58.

Ryburn, M. 2018. *Uncertain Citizenship: Everyday Practices of Bolivian Migrants in Chile*. Berkeley, CA: University of California Press.

Sassone, S. 1989. Migraciones limítrofes en la Argentina: Áreas de asentamiento y efectos geográficos. *Signos Universitarios: Ciencias Sociales y Geográficas (Revista de la Universidad del Salvador)* 3(15): 189.

Secretaría de Estado de Igualdad. 2016. *Denuncias Por Violencia de Género.* Madrid: Ministerio de la Presidencia, Relaciones con las Cortes e Igualdad.

Silvey, R. 2004a. Power, difference and mobility: Feminist advances in migration studies. *Progress in Human Geography* 28: 490–506.

Silvey, R. 2004b. Transnational domestication: State power and Indonesian migrant women in Saudi Arabia. *Political Geography* 23(3): 245–264.

Silvey, R. 2006. Geographies of gender and migration: Spatializing social difference. *International Migration Review* 40(1): 64–81.

Silvey, R. and R. Elmhirst. 2003. Engendering social capital: Women workers and rural–urban networks in Indonesia's crisis. *World Development* 31(5): 865–879.

Skar, S.L. 1993. The gendered dynamics of Quechua colonisation: Relations of centre and periphery in Peru. In *Migrant Women: Crossing Boundaries and Changing Identities*, ed. G. Buijs. Oxford: Berg.

Skeldon, R. 1997. *Migration and Development: A Global Perspective.* Harlow: Longman.

Valdés Echenique, T. and E. Gomariz Moraga. 1995. *Mujeres Latinoamericanas en Cifras: Tomo Comparativo.* Santiago: FLACSO.

Solé, C., S. Parella and A. Petroff. 2014. *Las Migraciones Bolivianas en la Encrucijada Interdisciplinar: Evolución, Cambios y Tendencias.* Barcelona: Universitat Autònoma de Barcelona.

Song, J. 2010. 'A room of one's own': The meaning of spatial autonomy for unmarried women in neoliberal South Korea. *Gender, Place & Culture* 17(2): 131–149.

Squires, J. 2008. Intersecting inequalities: Reflecting on the subjects and objects of equality. *The Political Quarterly* 79(1): 53–61.

Strunk, C. 2014. 'We are always thinking of our community': Bolivian hometown associations, networks of reciprocity, and indigeneity in Washington, DC. *Journal of Ethnic and Migration Studies* 40(11): 1697–1715.

Tapias, M. 2015. *Embodied Protests: Emotions and Women's Health in Bolivia.* Urbana, IL: University of Illinois Press.

Tapias, M. and X. Escandell. 2011. Not in the eyes of the beholder: Envy among Bolivian migrants in Spain. *International Migration* 49(6): 74–94.

Tizziani, A. 2011. De la movilidad ocupacional a las condiciones de trabajo. Algunas reflexiones en torno a diferentes carreras laborales dentro del servicio doméstico en la ciudad de Buenos Aires. *Trabajo y Sociedad* XV: 309–328.

Twigg, J. 2006. *The Body in Health and Social Care.* Basingstoke; New York: Palgrave Macmillan.

Vaa, M., S.E. Findley, A. Diallo and A. Diallo. 1989. The gift economy: a study of women migrants' survival strategies in a low-income Bamako neighborhood. *Labour, Capital and Society / Travail, Capital et Société* 22(2): 234–260.

Valentine, G. 2007. Theorizing and researching intersectionality: A challenge for feminist geography. *The Professional Geographer* 59(1): 10–21.

Van Hear, N. 2010. Theories of migration and social change. *Journal of Ethnic and Migration Studies* 36(10): 1531–1536.

Van Vleet, K.E. 2008. *Performing Kinship: Narrative, Gender, and the Intimacies of Power in the Andes.* Austin, TX: University of Texas Press.

Vertovec, S. 2001. Paper presented at workshop on 'Transnational Migration: Comparative Perspectives' WPTC-01-16, Oxford. http://www.transcomm.ox.ac.uk/working%20papers/Vertovec2.pdf

Voigt-Graf, C. 2004. Towards a geography of transnational spaces: Indian transnational communities in Australia. *Global Networks* 4(1): 25–49.

Vullnetari, J. and R. King. 2016. 'Washing men's feet': Gender, care and migration in Albania during and after communism. *Gender, Place & Culture* 23(2): 198–215.

Vullnetari, J. and R. King. 2008. 'Does your granny eat grass?' On mass migration, care drain and the fate of older people in rural Albania. *Global Networks* 8(2): 139–171.

Walsh, K. 2018. *Transnational Geographies of the Heart: Intimate Subjectivities in a Globalising City.* Oxford: Wiley Blackwell.

Wellman, B. 1999. *Networks in the Global Village: Life in Contemporary Communities.* Boulder, CO: Oxford: Westview Press.

West, C. and S. Fenstermaker. 1996. Doing difference. In *Race, Class and Gender: Common Bonds, Different Voices*, ed. E.N.-L. Chow, D. Wilkinson and M. Baca Zinn. London: Sage.

Whitesell, L. 2008. Y aquellos que se fueron: Retratos del éxodo boliviano. In *Desafiando la Globalización: Historias de la Experiencia Boliviana*, ed. J. Shultz and M. Crane Draper, 279–315. La Paz: Plural.

Wieringa, S. 1995. *Subversive Women: Historical Experiences of Gender and Resistance.* London: Zed Books.

Willis, K. and B.S.A. Yeoh. 2000. *Gender and Migration.* Cheltenham: Edward Elgar.

Wilson, A. 2012. Intimacy: A useful category of transnational analysis. In *The Global and the Intimate: Feminism in Our Time*, eds. G. Pratt and V. Rosner, 31–57. New York: Columbia University Press.

Winters, P., A. de Janvry and E. Sadoulet. 2001. Family and community networks in Mexico-U.S. migration. *The Journal of Human Resources* 36(1): 159–184.

Yarnall, K. and M. Price. 2010. Migration, development and a new rurality in the Valle Alto, Bolivia. *Journal of Latin American Geography* 9(1): 107–124.

Yarris, K.E. 2017. *Care Across Generations: Solidarity and Sacrifice in Transnational Families.* Stanford, CA: Stanford University Press.

Yeoh, B.S.A. 2006. Bifurcated labour: The unequal incorporation of transmigrants in Singapore. *Tijdschrift Voor Economische En Sociale Geografie* 97(1): 26–37.

Young, I.M. 1998. Harvey's complaint with race and gender struggles: A critical response. *Antipode* 30(1): 36–42.

Yuval-Davis, N. 2006. Intersectionality and feminist politics. *European Journal of Women's Studies* 13(3): 193–209.

Yuval-Davis, N. 2007. Intersectionality, citizenship and contemporary politics of belonging. *Critical Review of International Social and Political Philosophy* 10(4): 561–574.

Zack, N. 2005. *Inclusive Feminism: A Third Wave Theory of Women's Commonality.* Lanham, MD: Rowman and Littlefield.

Zhou, M.I.N. 1992. *Chinatown: The Socioeconomic Potential of an Urban Enclave.* Philadelphia, PA: Temple University Press.

Zunino, E. 1997. *Mujer Migrante Internacional en la Ciudad de Buenos Aires.* Buenos Aires: IOM and Secretaría de poblaciòn, CAREF.

Index

Note: Indicators in *italic* refer to figures and those in **bold** refer to tables.

Argentina 5, 7, 10–11, 70, 74, 91–93, 95–96, 126–127, 128
Autonomy 8, 18, 22, 39–40, 55, 76, 79–83, 147, 157, 160

Basch, L.G. 34, 35
Bolivia: and migration 2, 7, 9–10, 54–57; and remittances 1
Border 10, 11, 63–65, 126, 140
Bourdieu 38
Breadwinner 22, 45, 49, 53, 55, 89–90, 109–111, 129, 155, 158
Buenos Aires 5, 10, 13–16, 57, 49, 67, *72, 73,* 120, 132, 146

Care 19, 54, 90, 99, 113
Care worker 103–106
Caregiver 59
Chant, S. 24
Child protection legislation 123
Children 8, 49, 60, 113, 118–119, 121, 124–128, 131, 139
Class 6, 12–13, 22, 25, 27, 29, 47, 48, 161
Cleaner 106; *see also* domestic work
Cochabamba 36
Community 19–20, 36, 43–45, 46–47, 51–52, 70–75, 87, 155; *see also* transnational
Construction 97–98, 106–107
Country of origin 35, 38
Crenshaw, K. 12, 27, 31
Crisis (crises) 3–6, 11, 35, 47, 57, 104, 145; Argentinian 7, 11, 57, 63, 155; debt 35, 50; financial 11, 12, 79, 104, 106

Debt 9, 59, 76, 79–83, 97, 108, 116, 132, 154
Decision-making 18, 24, 55, 76, 78, 80–82, 159
Difference 25–26, 27–29
Disadvantage 13, 27, 29, 153
Domestic violence 16, 35, 50, 52, 75, 79, 82, 85–86, 144–148, 149, 158–159
Domestic work 95–97, 113

Education 11, 46, 49, 53, 82, 108, 116, 128, 144
Elderly 60, 103–106, 133
Elderly care 78, 103–106, 151–154
Emancipation 2, 36–39, 159–160
Emotion 138, 141, 142, 153–154
Empowerment 8, 21–22, 93, 110, 160; *see also* emancipation
Envy 60
Equality 31
Ethnicity 3, 12–13, 21, 25, 27, 31–32, 161
Ethnography 2, 138,
Externa 103; *see also* domestic work

Family 38, 59, 113–115, 121, 124, 129, 131, 135, 141, 150; and separations 118–120, 123, 157; and networks 157; extended 45, 68, **69**, 148; nuclear family 38, 39, 45, 158, 159
Family planning 49
Father(s) 36, 59, 75, 85, 115, 117, 123, 129, 139, 143, 158
Fear 94, 98, 116, 122
Femicide, *feminicidio* 147–148
Femininity 28, 158
Feminisation *see* migration

Feminism 27–28, 37; liberal 21, 25, 38; Western 25–26; black 12, 27–28, 33, 37; post-colonial 26, 37
Feminist geography 1, 39
Feminist: movement 12, 22, 27–28, 39, 160; scholarship 22, 37; geography 21–22, 26, 30, 36
Fieldwork 13–16, 36
Fraser 28, 32, 34

Garment sector/workshops 92, 93–95, 150
Gender 6–7, 21, 22–26, 33; and migration 13, 22–24, 35–38; gender-blind 25; gender ideology 45, 53, 55; gender relations 24, 37–38, 45, 52–54; gender roles 22, 26, 54, 111, 129, **130**, 158–159
Glick Schiller, N. 34, 35
Global care chains 114, 134
Grandmothers 114, 124, 134
Gregson, N. 25–26

Hochschild, A.R. 114
Hondagneu-Sotelo, P. 22, 55, 81–82, 124
House 4, 59, 67, **69**, 70–71, 77, 79, 81, 82, 83, 87, 108, *109*, 110, 116, 153; ownership 108, 111
Housework 89, 106, 114, 129, 158

Identity 39, 46–49, 51, 73; identity politics 27, 28
Indigenous 22, 48, 51
Inequality (inequalities) 21, 36, 113, 114, 158–159
Informal settlement 14–16, 51–52, 87, 116, 117, 120, 156
Informality 51; informal market 86; and childcare 122; and work 12, 99, 103–104
Intersectionality 2, 12–13, 26–29, 160–161; critiques of 29–33; origins 26–27; vectors 31
Intimacy 19, 137–138; 157; at work 151–154
Itinerant *see* migration

Journey 14, 18, 64, 68, 76–77, 114–116, 141
Justice 21, 28, 32, 34–35, 142

La Salada 14, 50
'Left behind': children 139–142; migrants' parents 113; spouse 22
Life story 9, 13, 15, 30, 43, 142, 160
Live-in 96, 203, 105; *see also* domestic work

Machismo 50, 54; machista 81
Masculinity 22, 28, 58, 61, 159
Masculinity index 56–57
McCall, L. 29–30
Methodology 2, 13–16, 29, 32, 160; and extended case method 2; and social networks 66; multi-sited 2, 3, 8, 13, 138
Migrant(s): primary 23; economic 23; men; 23; associational 23, 56–58
Migration: autonomous 39, 55, 75, 80–81, 159; complexity 8–9, 23, 43; internal 3, 7–8, 21, 39, 61, 90, 138, 155; itinerant 6–9, 137–138, 156–158; feminisation of 2, 7–8, 9–10, 34, 54–57, 61, 76, 102, 155; female 57; gender-selective 24; joint migration 76, 78–79; labour migration 116; permanent 77; regional 10, 57, 61, 67, 102, 135, 161; women-only migration 79–83; *see also* South-South
Mining town 3–5, 43, 45–51, 143, 145
Mobility 61; gendered 57; sequencing 76
Mother, mothers 55–56, 58–60, 85, 96, 100–101, 102, 105, 117, 143; single 29, 58, 59, 75, 96, 118–119, 135, 152,

Nash, J. 29–30, 32
Nations Unbound 35

Oppression 8, 12, 26–28, 30–32

Palliri 46, 89
Parenting 120–124; *see also* transnational
Parreñas, R. 56, 138
Patriarchy 27, 40, 159; patriarchal structures 1, 21, 24, 39, 55; patriarchal relations 7; patriarchal family 3, 7, 55; patriarchal norms 159
People, *gente* 3, 6, 77, 85, 151
Phizacklea, A.E. 23
Pregnancy 53

Race 12, 25–27, 30–31, 36
Racialisation 11
Radcliffe, S. 24, 52
Recognition 28, 32, 35
Redistribution 28, 32, 35
Remittances 1, 35, 107–108; and Bolivia 1
Rent 82, 83, 103, 120–122, 133; rented accommodation 116
Return/return migration 11, 60, 67, 72, 74, 77, 95, 108–111, 117–119, 127, 135, 157

Separatedness 40, 55, 85–86, 147
Sexual freedom 149–151
Sexual harassment 96–97
Sexual relations 49, 53, 147
Sexuality 32, 149
Social change 7, 21, 36, 37, 38, 40
Social mobility 6, 35, 51, 70, 81, 83–84, 87, 92, 96, 107–109, 117, 160
Social networks 12, 24, 44–45, 66–67, 156; and gender 66; and information 67; and accommodation 68–69; and labour market integration 69–70; and social capital 66, 68–70, 73–75; negative 73–74
Social relations 33–34, 38, 40
Social reproduction 113, 133, 157; *see also* care
Social transformation 36, 156; *see also* social change
South-South 1, 8, 113, 135; *see also* migration: regional
Spain 11–12, 50–57, 102–107, 146–148

Structural Adjustment Programme (policies) 3, 7, 10, 28, 35, 54
Survey 2, 13

Taxi 79, 109, *111*, 139
Teenage(r) 118, 127, 149; teenage pregnancies 53, 127
Temporality 77
Trading 98–102
Transnational mothering 124
Transnational: community 3, 9, 44, 145, 156; parenting 124–129
Transnationalism 7, 33–36, 114

Valentine, J. 30
Violence 63–64, 98, 120; *see also* domestic violence

Work 18–19; motivation for migration 83–84; mining 90
Working hours 74, 95, 98, 123

Yuval-Davis, N. 31